HIGH CURRENT, HIGH BRIGHTNESS, AND HIGH DUTY FACTOR ION INJECTORS

AIP CONFERENCE PROCEEDINGS 139

RITA G. LERNER
SERIES EDITOR

HIGH-CURRENT, HIGH-BRIGHTNESS, AND HIGH-DUTY FACTOR ION INJECTORS

LA JOLLA INSTITUTE 1985

EDITORS:

GEORGE H. GILLESPIE & YU-YUN KUO
PHYSICAL DYNAMICS, INC.

DENIS KEEFE
LAWRENCE BERKELEY LABORATORY

THOMAS P. WANGLER
LOS ALAMOS NATIONAL LABORATORY

AMERICAN INSTITUTE OF PHYSICS NEW YORK 1986

L.C. Catalog Card No. 86-70245
ISBN 0-88318-338-2
DOE CONF-8505164

Printed in the United States of America

TABLE OF CONTENTS

PREFACE

One of the important frontier areas in accelerator physics today is the search for ever better ways of simultaneously achieving higher beam currents with very low emittance, and preserving both properties throughout the acceleration process. The workshop on High-Current, High-Brightness, and High-Duty Factor Ion Injectors held in San Diego, May 21–23, 1985, focused on the basic and applied physics problems encountered with intense ion beams in the low velocity end of a linac. A primary objective of this workshop was to review in depth the different approaches to intense ion injectors, and the phenomena associated with high-current and high-brightness beams. A second objective was to provide a forum for discussion for those people active in this specialized field, and thereby stimulate suggestions and recommendations regarding future work in these areas. We believe the workshop was very successful in regard to these objectives.

The subject material was grouped under five main headings: diagnostics, theory (including simulation), high-current beam transport, cw radio frequency quadrupole accelerators, and multiple beams (including funneling). Following introductory reviews by experts in each of these areas, participants divided themselves among five corresponding working groups. The review papers and the working group summaries and conclusions form the main body of this Proceedings. In addition, four contributed reports on new work presented at the workshop are also included.

The workshop and publication of the Proceedings were supported by the U.S. Department of Defense and the U.S. Department of Energy to whom we express our gratitude. We are especially grateful to Mr. Milton C. Hawie of the Ballistic Missile Defense Advanced Technology Center in Huntsville and Drs. David F. Sutter and Terry Godlove of the Office of Energy Research, in DOE Germantown, for their advice and encouragement.

This meeting was very successful and well-received by all in attendance—many participants urging that it be held bi-annually in the future. Many people and organizations contributed to the success of this workshop and we acknowledge at least some of them here. The organizing committee is especially grateful to the La Jolla Institute (LJI) for its willingness to host the meeting and to apply its considerable expertise to the administrative chores. We wish to thank Dr. Adolf Hochstim, President of LJI, for his support and commitment to the workshop, without which this endeavor would not have been possible. Physical Dynamics, Inc. (PDI) is due a double thanks for providing the local committee and the conference secretarial support. Last, but certainly not least, special thanks are due to Marcellite Broadhurst (PDI) who handled the meeting arrangements with the Viscount Hotel on Harbor Island, to Betsey Besaw (PDI) who assisted her during the meeting, to Grace Pitts (LJI) who handled the advance registrations and room reservations, and to Rosalie Rocher (LJI) who handled the Proceedings manuscripts.

George H. Gillespie and Yu-Yun Kuo
Physical Dynamics, Inc.
Denis Keefe,
Lawrence Berkeley Laboratory
Thomas P. Wangler,
Los Alamos National Laboratory

HIGH-BRIGHTNESS, HIGH-CURRENT ION SOURCES*

Roderich Keller,† MS H816
Los Alamos National Laboratory, Los Alamos, NM 87545

ABSTRACT

Several existing high-current dc ion sources are presented that yield positive ions and are based on gas discharges. Their common features are then outlined: extraction systems, required plasma qualities, and the production of ions from solid materials. Finally, from well-established laws, scaling rules for high-brightness ion beams are derived taking one specific case to determine the numerical constants. These rules predict that the current values of transportable beams scale proportionally to the 3/2 power of the extraction voltage, whereas the brightness strongly decreases with rising voltage.

INTRODUCTION

The definition of high ion current is linked to space-charge phenomena, but there is no sharp threshold where these phenomena begin. The effects depend on current density, beam energy, and also on the beam quality involved. Nevertheless, many sources delivering beams between some milliamperes and tens of amperes ion current can be grouped together as high-current sources because they exhibit several common features.

This paper presents a variety of dc high-current ion sources suitable for linear accelerators and emphasizes their common aspects, concentrating on plasma sources for positive ions. Scaling rules for brightness dependence on extraction voltage are derived from several well-established laws, independently of any particular source type. The numerical factors in these rules, however, are based on the performance of one existing source only. A comparison of current values actually obtained by many other sources with the current scaling law here proposed seems to justify this derivation.

INDIVIDUAL SOURCE TYPES

Duopigatron

The source with which most of the pioneering work has been accomplished is the duopigatron.[1] Essentially, it is based on a two-stage discharge between a hot cathode, an intermediate electrode, and an anode, see Fig. 1. The plasma is guided by a

*Work done at Darmstadt.
†Visiting scientist from GSI, Darmstadt, West Germany.

carefully tailored axial magnetic field through the anode channel into a third chamber, at the end of which the electrons are reflected back toward the intermediate electrode. With more sophisticated models,[2] additional electrodes are applied to control the expanding plasma. The duopigatron offers good gas and power efficiencies, but the discharge suffers from strong oscillations if the electrode geometry is not optimized or the optimum operation conditions are not applied.

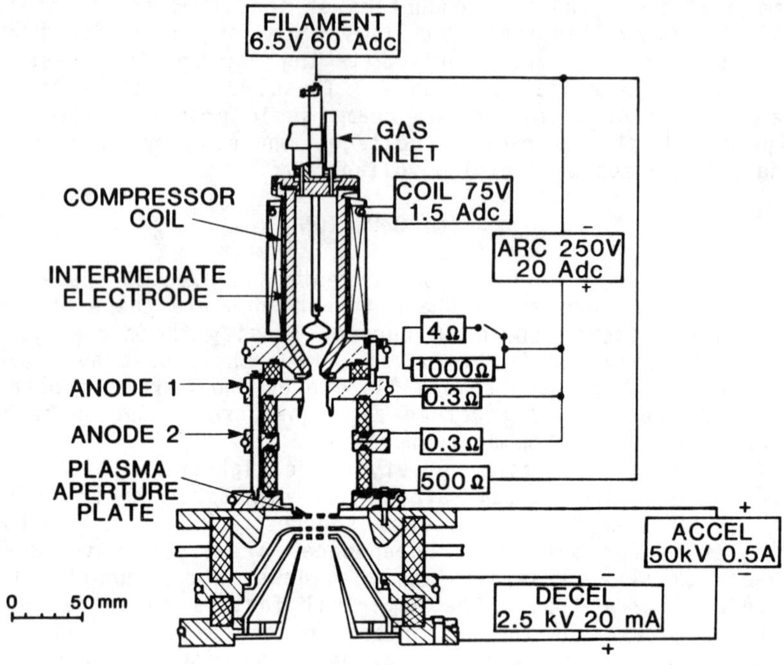

Fig. 1. Duopigatron ion source.[3]

"Penning" Source

An efficient as well as simple source[4] of the Penning type--not to be mistaken for a PIG source of highly charged ions--effectively resembles the second half of a duopigatron, with one difference: the electrons sustaining the discharge are directly emitted by a filament, without need for a first discharge stage, see Fig. 2. The axial magnetic field is generated by permanent magnets placed around the cylindrical anode and expands from the cathode plane towards the outlet electrode (called "focus electrode" in Fig. 2).

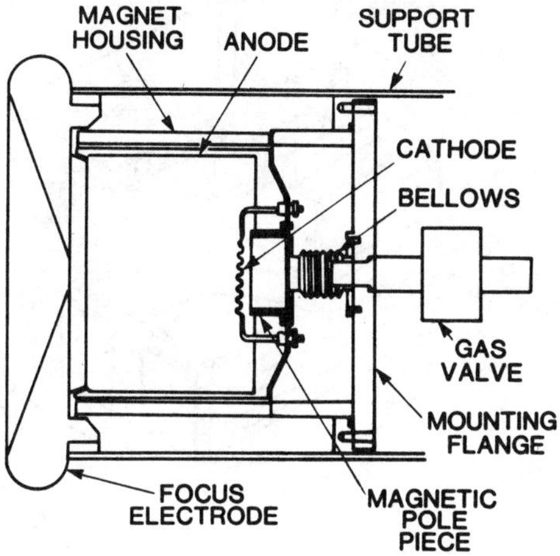

Fig. 2. "Penning" high-current ion source.[4]

Multicusp Sources

With the availability of strong permanent magnets, multicusp discharges are more and more being used to create the plasma needed for high-current sources. The discharge vessel is surrounded by magnets with alternating polarity, creating a minimum-B configuration that yields a stable plasma confinement over many decades of plasma density.[5] The rear source side may be protected by magnets (Fig. 3) or biased,[7] depending mostly on technical considerations. The multicusp sources, in general, offer the highest versatility whenever a wide range of ion species, beam intensity, and beam quality is to be provided by one basic source type and when other features like gas- and power-efficiency or ionization degree are of minor importance.

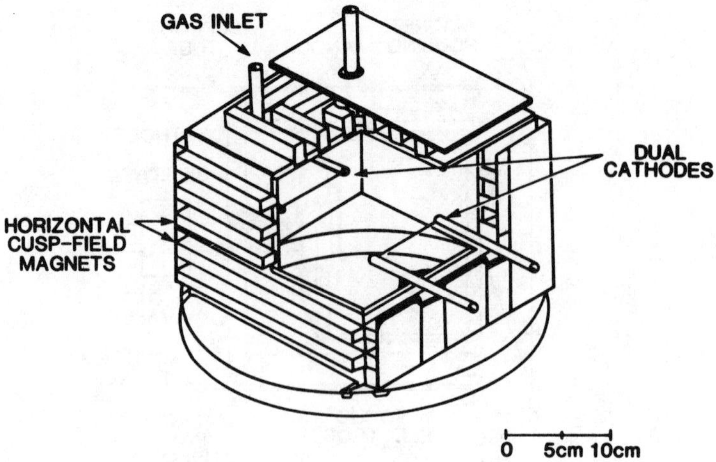

Fig. 3. Multicusp ion source.[6] Ions are extracted towards the bottom.

Monocusp Source

A rather new source type[8] combines the qualities of the "Penning" and cusp sources, using a single, ring-shaped magnet around the discharge vessel (Fig. 4). This source design results in excellent gas efficiency because plasma losses are very low, but the design somewhat limits the useful plasma cross section near the outlet electrode because of the radial variation of the plasma density.

High-Frequency Sources

Sources using discharges between polarized electrodes always suffer from erosion, mostly of the cathodes. High-frequency discharge sources are free of this defect, but one must consider whether or not this advantage pays for the more complicated equipment involved. The frequencies used generally range from several hundreds of kilohertz to some tens of gigahertz. The plasma vessels can be made from insulators or metal; in the latter case, they are often formed as a resonator. Magnetic fields can be used to confine the discharge or, in the case of microwaves, to produce electron cyclotron resonance (ECR) effects. An example of such sources,[9] equipped with a magnetic multicusp field, is shown in Fig. 5.

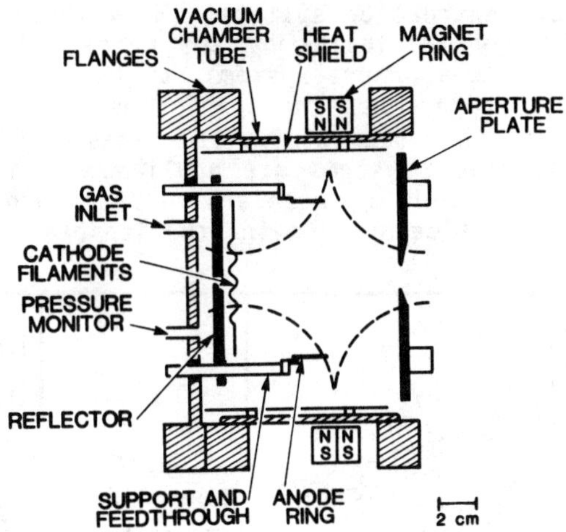

Fig. 4. Monocusp ion source.[8] Broken: magnetic field lines.

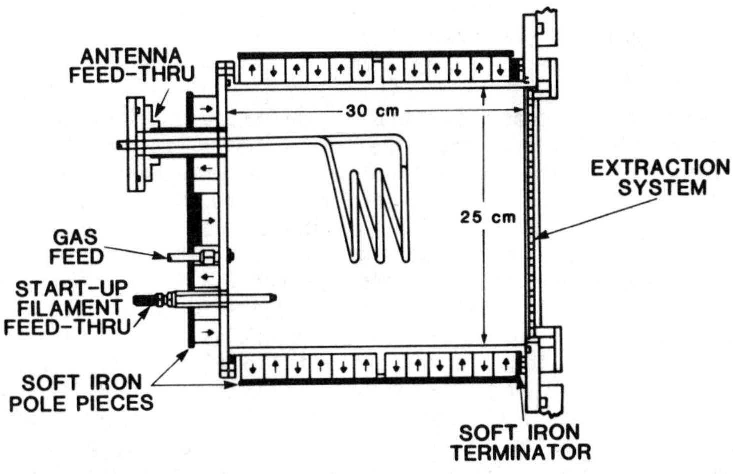

Fig. 5. High-frequency ion source (2 MHz) with multicusp.[9]

COMMON FEATURES OF HIGH-CURRENT SOURCES

Extraction Systems

All source types presented above are equipped with large-area outlet apertures; most of them can be used with single- or multiple-aperture extraction systems. With the latter, the total beam current simply multiplies by the number of apertures (as compared to a single-aperture system) but the overall emittance may increase drastically because the spacing between the apertures has to be added to the beamlet waist sizes.[10] Commonly, accel/decel extraction systems are applied, see Fig. 6, to avoid compensating electrons being accelerated back into the source, thus destroying its rear side and leaving the ion beam uncompensated.

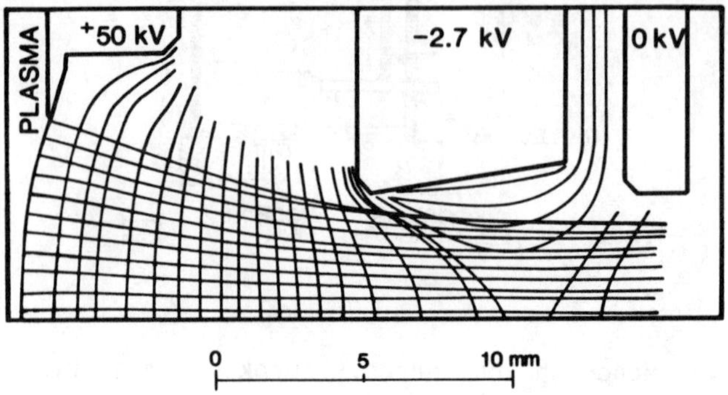

Fig. 6. Accel/decel extraction system[11] (half section, computer simulation of ion trajectories).

The most important feature of an extraction system is the contour of the outlet aperture. The point is to form the equipotentials in the extraction gap in a way that aberrations are reduced, taking into account also the space-charge action of the resulting beam. This aperture shaping must always be paid for in terms of extraction field strength. As a general rule, it is convenient to reduce the thickness of the outlet electrode near its aperture rim as much as the cooling requirements permit. The outer contour then should be bell-shaped, but no further general hint can be given, the individual needs being too different from one another.

A quite promising technique consists of dividing the extraction gap into two gaps (Fig. 7). Thus, the total width can be shortened, allowing higher fields to be applied, because of the nonlinear relation between necessary gap width and voltage, see below. The major benefit, however, is the possibility of using the voltage applied to the inserted electrode as an additional free parameter to influence the shape of the equipotentials. This technique seems to be convenient at extraction voltages above 50 kV, but in general for single-aperture systems only.

Plasma Qualities

The first condition imposed on an ion source plasma is that its density must match the applied extraction field to yield a low-divergence beam. Uniform ion emission is required across the

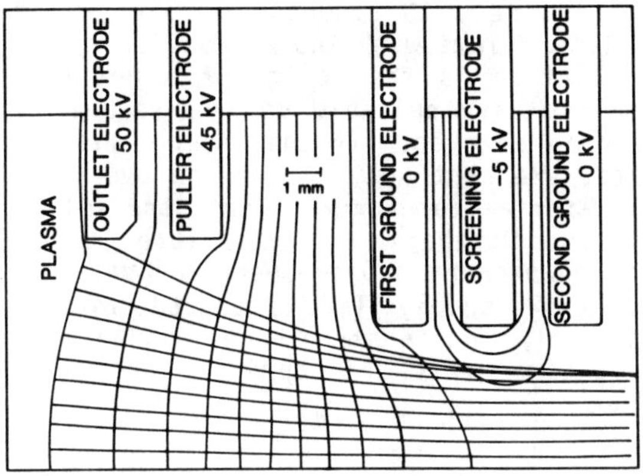

Fig. 7. Two-gap extraction system.[10] Here the screening electrode is enclosed between two ground electrodes to improve the system reliability, following Ref. 6.

whole outlet area, even if only one aperture is used. In the case of plasma sources, this requirement implies that the plasma column must be much wider than the outlet aperture.

Plasma oscillations, especially in the 100-kHz range, must be minimized to avoid destroying the space-charge compensation of the beam. To yield optimum beam quality, the transverse ion energy must be low, typically some tenths of electron volts.

These quiet, cold plasmas usually are not suited for producing highly charged ions, but considerable fractions of doubly charged ions have been found,[12] in some cases. Multicusp sources with their large regime of operation parameters can quite easily be tuned to produce doubly and higher charged ions because they can operate stably with relatively high voltage and low gas pressure, the most suited conditions for this purpose. It must be noted, however, that there always exists a distribution over different ion charge states. In consequence, individual higher charge states can be extracted from a high-current source with less than one-third the current obtained when producing singly charged ions only.

Ions from Solids

Although the generation of ions from gases is simple enough and mostly exhibits lifetime problems, solids address quite another class of difficulties in source development and operation. The use of volatile compounds with conventional sources brings three disadvantages: the total extracted ion current must be shared between the different ion species, including ionic molecules; the compound usually contains an aggressive component, like halogens, which reduces the source's lifetime; and the free atoms of the solid element are likely to condense on the cold source walls, which may lead to problems with the extraction system.

A better solution is the design of a hot source (Fig. 8) because many solids have suitable vapor-pressure curves that they can be processed in an oven. If the inner source walls, including the outlet plate, are kept hotter than the oven, no recondensation of the feeding material can occur. Thus, the solid material can be used like a gas, with the oven temperature replacing the gas pressure as an independent source-operation parameter. In another approach,[13] a metal supply is introduced into the plasma expansion chamber of a duopigatron. An auxiliary gas maintains the discharge; the plasma heats the supply to provide free atoms and ionizes the atoms.

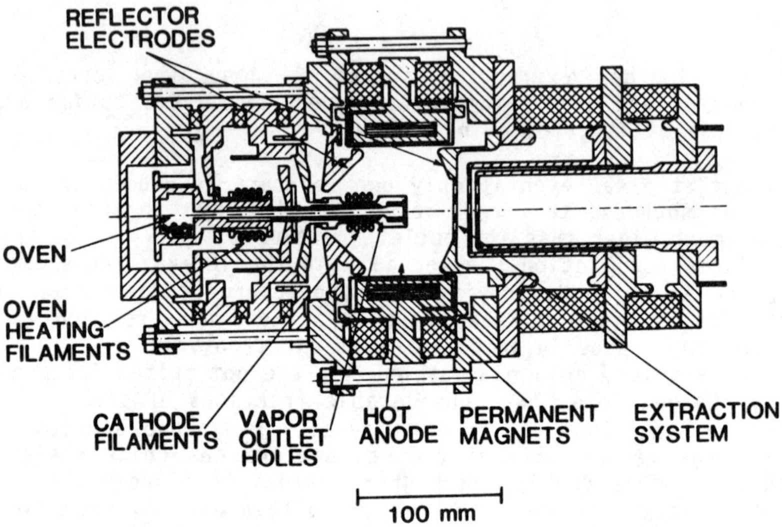

Fig. 8. Hot running multicusp ion source,[12] for use with nonvolatile elements.

With such sources, solid material will always be deposited on the relatively cold extraction electrodes. This layer cannot be re-evaporated because a discharge would arise in the extraction gap; thus, larger gaps than those used with gas sources must be applied, allowing less ion current to be drawn out of the source.

SCALING RULES FOR HIGH-BRIGHTNESS BEAMS

The quality of an ion beam is commonly judged by its brightness value, as defined in the following paragraph. Frequently some misunderstanding occurs concerning this quantity.

In the accelerator community, people are familiar with the Liouville theorem which, when applicable, states that for a given beam the normalized emittance remains constant. This rule sometimes induces one to expect the maximum normalized brightness of an ion source to be a constant, but that is completely wrong.

Beam-Quality Definitions

In this paper, the following definitions will be used: I, ion current emitted by the ion source; I_{tr}, transported part of the total current; $I_n = I\ (A/\zeta)^{1/2}$, normalized current (proton equivalent of any ion beam of atomic mass A and charge ζ); U, effective extraction voltage, defining the beam energy; $P = I_n/U^{3/2}$, perveance; d, extraction gap width; r, outlet aperture radius; $S = r\ /\ d$, aspect ratio of the system; r_{max}, beam-waist radius.

The measured emittance is $\varepsilon = A_E/\pi$ (here called absolute emittance), with A_E, area of the emittance figure; $\varepsilon_n = \beta\gamma\varepsilon$, normalized emittance; β and γ, usual relativistic parameters. For ion sources and low-energy beams, $\gamma = 1$ is always fulfilled, and the velocity can be evaluated from $\beta = 1.46 \times 10^{-3}\ (\zeta\ U/A)^{1/2}$.

A common definition for the brightness of a cylindrical beam reads: $B = I_{tr}/\varepsilon_n^2$. It seems indicated, however, to distinguish between this quantity (better named $B_{\varepsilon n}$, "emittance-normalized" brightness) and a newly introduced "current-normalized" brightness B_{cn}, defined as $B_{cn} = I_{tr,n}/\varepsilon^2$, using the quantities normalized current and absolute emittance.

Voltage Dependence of Beam Parameters

The following discussion is intended for conventional plasma ion sources, and only one round aperture is considered.

In a thorough experimental study of a duopigatron,[14] it was found that the actually transported ion currents, with low divergence half-angle, scale as $I_{tr,n} = P^*S^2U^{3/2}/(1 + aS^2)$, see Fig. 9a, rather than $I \propto S^2\ U^{3/2}$ as suggested by the Child/Langmuir law[15] for the total extracted currents. The factors a (aberration factor) and P^* (low-S perveance) depend upon the divergence half-angle α_0 accepted by the transport system, and also upon the chosen extraction electrode geometry. The quoted study states $a \simeq 3$. For a more refined two-gap extraction system,[10] $a = 1.7$ and $P^* = 6 \times 10^{-8}$ $(A/V^{3/2})$ give the best fit to the current values measured within $\alpha_0 = 20$ mrad, see Fig. 9b.

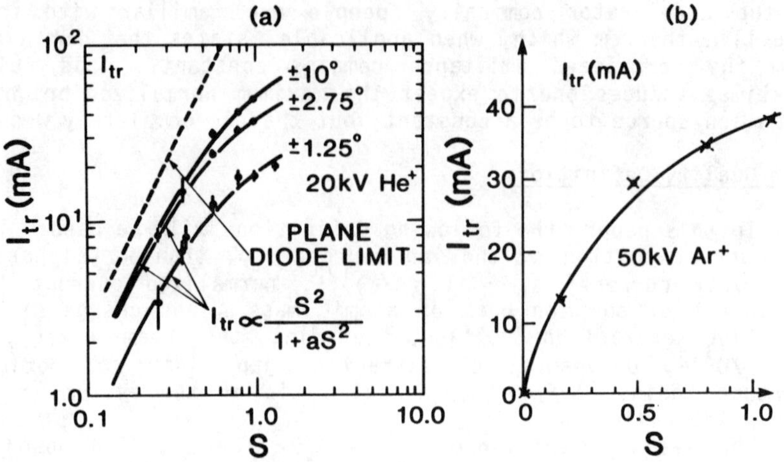

Fig. 9. Transported ion currents I_{tr} versus aspect ratio S of extraction system: (a) triode[14] (log-log plot), (b) pentode[10] (linear plot).

The existence of a saturation limit for the transportable current with increasing aspect ratio S strongly recommends maintaining $S \simeq 0.5$, when designing extraction systems, because the spilled-beam part extracted out of larger openings causes quite unfavorable effects. Only sophisticated geometries,[11] (Fig. 6) or two-gap systems permit the application of values $S \simeq 1$ without losing too much in reliability of the extraction system.

Taking the values P^* and a of the above-quoted two-gap system[10] as reference and allowing for S = 1, one can conclude at once that the maximum normalized current to be obtained from one round aperture and transported within 20-mrad divergence half-angle is a function of the voltage only and (in convenient units) amounts to

$$I_{tr,n} = 7.03 \times 10^{-4} U^{3/2} \ (\mathrm{A/kV^{3/2}}) \ (\text{see Fig. 10}) \quad .$$

In the usual case where the emittance figure of a beam is contained within an ellipse, the beam emittance in the plane of its waist is given by the product $\varepsilon = \alpha_0 \, r_{max}$, where α_0 is the maximum angle of the trajectories crossing the beam axis in this plane. It has often been found (and confirmed for the two-gap system[10] by computer simulations and emittance measurements) that for cylindrical high-brightness beams, the waist has one-half the extraction aperture width: $r_{max} = r/2$. Considering that 20 mrad was the acceptance half-angle for the system under discussion, it results that now the absolute emittance of the transported beam is only a function of r (which is equal to d) and amounts to $\varepsilon = 0.5 \times 20\ r = 10\ d$ (mrad).

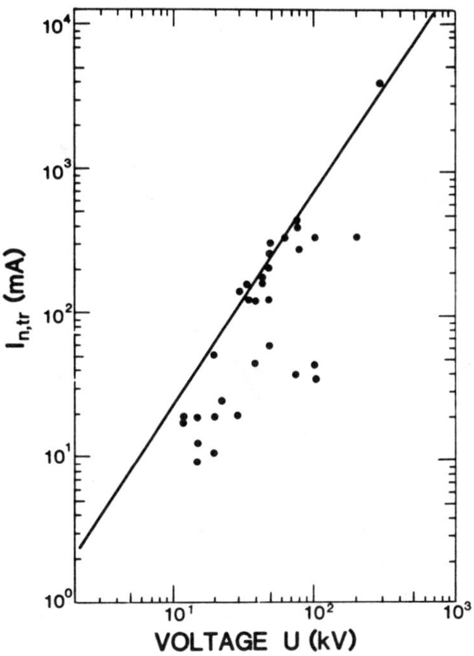

Fig. 10. Voltage dependence of normalized transported current $I_{n,tr}$. The points mark the $I_{n,tr}$-values of actual sources, calculated for one emitting aperture only. Current values that lie significantly below the marked limit mostly belong to neutral injection experiments, where conditions other than maximum transported current are dominant.

Regarding the minimum d-value, collected data of many existing high-current sources[16] show that the implicit Kilpatrick law[17] marks a close lower limit for the gap widths through two decades of voltages (Fig. 11) and that a reasonably cautious empirical limit is given by the explicit formula

$d = 0.01414\ U^{3/2}$ (mm/kV$^{3/2}$), permitting 50 kV for a 5-mm gap.

Taking an optimized system with S = 1 then yields

$r = d = 0.01414\ U^{3/2}$ (mm/kV$^{3/2}$) and, in consequence, for the absolute emittance $\varepsilon = 0.1414\ U^{3/2}$ (mm mrad/kV$^{3/2}$).

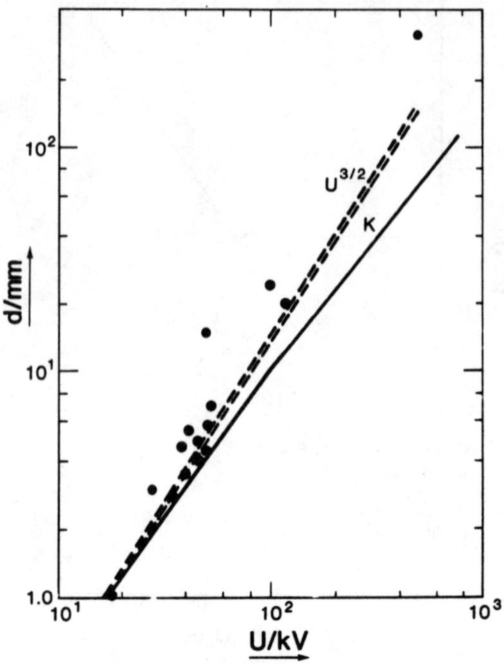

Fig. 11. Breakdown limits for extraction gaps. K, Kilpatrick law,[17] $U^{3/2}$, empirical limit.

For the normalized emittance, it directly follows that

$$\varepsilon_n = 2.06 \times 10^{-4}\ (A/\zeta)^{-1/2}\ U^2\ (\text{mm mrad/kV}^2).$$

The conclusion to be drawn for the dependence of the brightness values from the extraction voltage looks quite unfamiliar; the emittance-normalized (commonly used) as well as the newly introduced current-normalized brightness quantities strongly decrease with increasing voltage (Fig. 12):

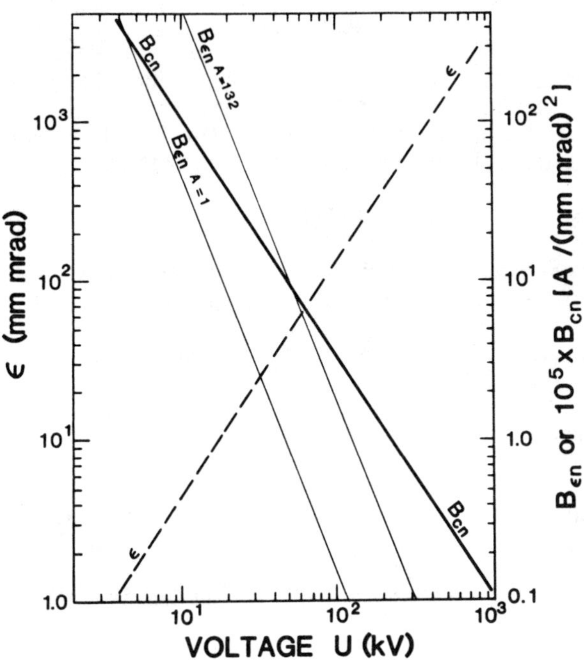

Fig. 12. Voltage dependence of absolute emittance ϵ, emittance-normalized brightness $B_{\epsilon n}$, and current-normalized brightness B_{cn}, calculated according to the formulae derived in the text.

$$B_{\epsilon n} = 1.65 \times 10^4 \ (A/\zeta)^{1/2} \ U^{-5/2} \left[A \ kV^{5/2}/(mm \ mrad)^2\right]$$

and

$$B_{cn} = 3.52 \times 10^{-2} \ U^{-3/2} \ \left[A \ kV^{3/2}/(mm \ mrad)^2\right] \ .$$

Of course this scheme does not mean that the derived numerical factors were absolute constants. The absolute brightness values may vary considerably, depending on the emittance definition (fraction, rms value), the measurement technique, and also on the assumed acceptance half-angle; however, the validity of the tendencies here developed for individual sources appears to be well confirmed, and (much more importantly) over 30 different collected source data fit closely into the ion-current scheme (Fig. 10), including a contact-ionization source for which the above discussion does not apply at all.

The proposed scaling rules may therefore be regarded at least as a standard reference to which any high-current ion source can be compared, to judge its performance.

ACKNOWLEDGMENTS

Thanks are due to Los Alamos National Laboratory, particularly to S. Schriber, L. Taylor, and P. Allison, for critically examining and editing this manuscript, and to E. Vigil and D. Martinez for their excellent work in redrawing the illustrations.

REFERENCES

1. R. A. Demirkhanov, H. Fröhlich, V. V. Kursanov, T. I. Gutkin, Brookhaven National Laboratory report BNL-767, 224 (1962).

2. R. C. Davis, O. B. Morgan, L. D. Stewart, W. L. Stirling, Rev. Sci. Inst. 43, 278 (1972).

3. M. R. Shubaly, Inst. Phys. Conf., Ser. 54, Bristol, p. 333 (1980).

4. R. P. Vahrenkamp, R. L. Seliger, IEEE Trans. Nucl. Sci. 26, 3101 (1979).

5. K. N. Leung, N. Hershkowitz, K. R. MacKenzie, Phys. Fluids 19, 1045 (1976).

6. J. D. Schneider, H. L. Rutkowski, E. A. Meyer, D.D. Armstrong, B. A. Sherwood, L. L. Catlin, Brookhaven National Laboratory report BNL-51134, 457 (1979).

7. R. Keller, F. Nöhmayer, P. Spädtke, M. H. Schönenberg, Vacuum 34, 31 (1984).

8. J. P. Brainard, J. P. O'Hagan, Rev. Sci. Instr., 54, 1497 (1983).

9. V. V. Fosnight, W. F. Di Vergilio, report EDB 700205, TRW, Redondo Beach, California, USA (1983).

10. R. Keller, P. Spädtke, K. Hofmann, Springer Ser. Electrophys. 11, Heidelberg, p. 69 (1983).

11. B. Piosczyk, Los Alamos National Laboratory report LA-9234-C (February 1982).

12. R. Keller, P. Spädtke, F. Nöhmayer, Int. Ion Engin. Congr., Kyoto, p. 25 (1983) and K. N. Leung and R. V. Pyle, submitted to IEEE Int. Conf. on Plasma Sci., Pittsburg (1985).

13. P. G. Weber, Rev. Sci. Instr., 54, 1506 (1983).

14. J. R. Coupland, T. S. Green, D. P. Hammond, A. C. Riviere, Rev. Sci. Instr., 44, 1258 (1973).

15. C. D. Child, Phys. Rev., 32, 492 (1921).

16. R. Keller, Gesellschaft für Schwerionenforschung report GSI-82-8, Darmstadt, p. 87 (1982).

17. W. D. Kilpatrick, Rev. Sci. Instrum., 2, 824 (1957).

CW RFQ DEVELOPMENT*

S. O. Schriber, AT-DO, MS H811
Los Alamos National Laboratory, Los Alamos, NM 87545

ABSTRACT

A review of research and development related to fabricating and operating radio-frequency quadrupole (RFQ) structures at 100% duty cycle [continuous wave (cw)] is presented, with emphasis on work at the Los Alamos National Laboratory, the Chalk River Nuclear Laboratories, and the University of Frankfurt. Activities in other areas that have an impact on operating cw RFQ systems will be highlighted.

INTRODUCTION

A great amount of development activity followed the 1977 introduction of the RFQ principle from the USSR. The RFQ concept was initially developed by Kapchinskii and Teplyakov[1] in 1969, was verified in a prototype (double-H) cavity by Golosai et al.[2] in 1974, and was demonstrated as an injector for the URAL-30 linac by Gorshkov et al.[3] in 1977. Good performance was achieved in the two referenced studies, showing suitability of an RFQ for bunching and accelerating low-energy ion beams with good efficiency--both rf power and percentage beam capture.

Initial work outside the USSR was championed by the Los Alamos National Laboratory with a demonstration of the suitability of the four-vane geometry[4] and with the verification of proton beam acceleration on Valentine's Day 1980. The proof-of-principle RFQ structure[5] was the precursor to the first design and dedicated study effort of a cw RFQ. Development of the RFQ was both timely and fortuitous in that solutions were being sought for the initial beam formation of the Fusion Materials Irradiation Test Facility (FMIT)[6] at Los Alamos and the Electronuclear Breeder[7] at Chalk River. Both of these high-power systems were being designed to accelerate high-current cw ion beams--100 mA D^+ and 300 mA H^+ respectively. In addition, the cw beam formation had to be efficient (rf power and capture) with small emittance growth and little halo formation, because beam spill along the length of accelerator had to be minimized.

The choice or requirement of cw structures puts extra constraints on the design, fabrication, and operation of a device, as compared to the normal practices and the technological base for pulsed systems. Particular attention has to be given to

- cooling all rf surfaces
- rf penetrations
- thermal conduction at all mechanical joints and interfaces
- frequency control methods

*Work supported by US Department of Energy.

- surface imperfections at high rf electric field locations
- vacuum requirements for lost beam
- lost-beam interaction products and surface outgassing
- induced radiation
- multipactor
- x-ray fields and shielding
- ultraviolet and visible light production
- ceramic charging
- rf window locations
- thermal expansion and possible distortion
- vibration
- surface erosion
- diagnostics
- movable rf joints
- beam-dynamics-related mechanical tolerances
- selection of materials and joining techniques
- beam transients
- reverse power and rf isolation
- alignment
- selection of beam bore hole size
- control and monitoring systems
- safety devices and fast beam aborts
- beam dumps
- transport elements

The following discussion is based on experiences and insights that have been gleaned over the past several years in association with cw structures.

RFQ DEVELOPMENT

A number of Laboratories and Institutions have been investing effort and developing the RFQ structure in its four main geometries--four-vane, four-rod, split-coax, and spiral-resonator. Although most of the activities have been for pulsed or low duty-factor systems, some of the research and development is directly applicable to a cw device. These aspects will be discussed in a later section covering miscellaneous items.

Los Alamos National Laboratory

The 80-MHz FMIT[6] RFQ employs a coaxial manifold to couple rf power via eight slots to the 3.9-m RFQ structure. The manifold is driven in the TEM coaxial mode by an off-center (longitudinally) coupling loop. Short-length coaxial cavities are employed at each end of the manifold to provide a region for shielding cooling line connections to the RFQ outer wall and vanes, and to produce a simplified rf shorting joint at either end of the effective volume of the manifold. Other holes in the manifold were used for viewing ports, piston tuner ports, and vacuum connections.

The vanes of the four-vane RFQ geometry are copper-plated mild steel. The rf joints to the outer wall are made by gold-plated copper seals. A copper-plated split cylinder completely encircles the vanes, providing a flexible joint between each vane and the vane base mount. In this manner, vanes can be finely adjusted to the correct position following an initial gross alignment when each is bolted to the outer wall. Several rf pickup probes are mounted along the side of the vanes in each quadrant.

Initial operation[8-10] was plagued by overheating and multipactor problems, primarily in the manifold. The multipactor was cured by TiN plating of the entire vacuum envelope and by a redesign of the rf joint between the short-length coaxial cavity and the coaxial manifold. Overheating problems were slowly overcome by a redesign of the covers for the vane cooling lines, improving rf mechanical joints and removal of end-vane tuner components that were overheating badly. The flexible, split-cylinder joint also experienced some overheating that led to thermal runaway. Welding of solid-copper patches and improved cooling overcame this difficulty. The two most important lessons learned during initial operation were to properly cool everything and to consider every aspect of possible rf coupling from one area to another.

With all of the added improvements and modifications, the FMIT RFQ now operates reliably and on demand. Approximately 30 minutes are required for the system to reach equilibrium, after which extensive beam tests can be conducted by the experimental team employing needed automatic control systems and noninterceptive diagnostics.

Beam measurements show good agreement between PARMTEQ beam-dynamics simulations and experimental results. A few discrepancies have been noted, including a sudden shift in beam position at very high power levels and a different orientation of the y-y' phase space. The former is probably related to harmonic generation from the rf system at very high power. The latter may be related to an additional half-cell effect at the exit of the RFQ--the region between the vane ends and the smaller beam exit aperture. In addition, measurements involving an adjustment of injector parameters indicate that the front plate of the RFQ has become slightly magnetized. The end-plate situation has a noticeable effect on beam-current transmission as a function of excitation of the solenoid in proximity to the RFQ entrance.

The cw H_2^+ beam (currently limited to 20 mA by administrative control) is a powerful beam and has to be handled carefully, otherwise holes can be burned in beam pipes and organic vacuum gaskets can be ruined very quickly. Visual observation of the beam (by observing residual gas ionization) is a definite asset and a diagnostic necessity for any cw structure.

Chalk River Nuclear Laboratories

Early cw experience with 800-MHz, cw, coupled-cavity structures;[11] cw electron-beam acceleration;[12] 270-MHz drift-tube-linac structures;[13,14] and cw proton-beam acceleration[15] demonstrated the need for good diagnostics, safe and effective control systems, and adequate cooling on all rf surfaces (with careful design of any rf or thermal mechanical joint). High-field conditioning procedures for structures were also demonstrated. The procedures employed a combination of pulsed and cw excitation while carefully observing vacuum, breakdown rate, and reverse rf power. Beam-dynamics considerations and lost beam factors suggested that beam bore holes in structures should be at least 7 to 10 times the beam rms size.

The 0.4-m, 270-MHz, four-vane RFQ[16] that was operated cw for several months provided much useful information on design and operation of RFQ structures. Because the vanes were made from solid OFHC copper, no visible erosion was noted on the tips after many hours of high-voltage stress tests--unlike the erosion that is prevalent with copper-plated surfaces. Vane voltage was determined by measuring the end-point of the x rays created by the electrons that are stopped on the vane tips after traversing the vane gap. Visible light from the microdischarge "glow points" was useful in assessing local surface characteristics and the onset of breakdown. Ultraclean techniques, similar to that used for superconducting structures, should be employed on room-temperature structures to minimize the number of "glow points" and to improve operation capabilities in terms of breakdown rate and the breakdown voltage level.

Design is under way for a high-power, cw, four-vane RFQ[17] to demonstrate cw operating characteristics, beam transport, beam limitations with either multiple or single beamlets, and the accuracy and validity of different beam-dynamics simulation codes. The accelerator, called RFQ1, will operate at 270 MHz with vane electric fields ranging from 1.5 to 2 Kilpatrick and will require up to 200 kW of rf power for the 1.5-m length vanes (about 40 kW additional is necessary for the 600-keV output beam). Some experiments on beam loading will be part of the experimental program, especially for the lower rf excitation levels at which point 30% beam loadings should be possible.

The cw resonant-load facility[18] has been demonstrated to be a valuable asset--as a similar facility would be to any cw program. With this empty right-circular cylinder, that was constructed with multiple entry ports, different components of a cw structure can be tested individually at high power to determine their suitability. This facility was used to test rf coupling loops, rf windows, piston tuners, drift tubes (for a drift-tube linac) and the proposed rf vane-to-outer-wall joint for RFQ1.

Based on high-power tests and several overall design iterations, the RFQ1 concept[19] has some very interesting features. The vanes are replaceable because of the incorporation of a vane to outer wall joint that is similar to standard conflat vacuum joints. The copper seal has been successfully high-power tested on the resonant-load facility. Vane tips are made from solid OFHC copper that has been brazed to the vane mount. Brazing such a joint with Pacusil 5 has been successfully demonstrated for a 1.5-m-length test vane.

Vacuum ports have been designed[20] to minimize rf leakage and to maintain a good rf quality factor by ensuring that the ratio of the hole length to its diameter is not less than 1.3. Flood cooling, close attention to thermal expansion, vane coupling rings (VCRs) at the vane ends, and good clean-room techniques should ensure that this accelerator will be commissioned quickly and will demonstrate good cw performance in late 1986 as scheduled.

University of Frankfurt

Studies of surface breakdown[21] and operation of cw four-rod structures[22] has been providing useful information to the whole accelerator community. Surface breakdown studies have shown that with shorter pulse lengths one can support higher voltages between the vane tips. Higher fields are important because they lead to higher accelerating and focusing fields, which will increase the beam-current capability of a particular structure. As long as vacuum pressures are less than 10^{-5} torr, there should be no noticeable change in the breakdown voltage level. The addition of cesium vapor had a deleterious effect (almost halved) on vane voltage capabilities. Tests with cylindrical electrodes showed that standard OFHC copper is better than chromium copper (0.75% Cr) by about 10% in voltage stand-off capabilities.

Studies on electrode shape[23] showed that simpler electrode profiles can be used without affecting the beam dynamics in a significant manner.

Design, fabrication, and operational beam tests of the four-rod, spiral-resonator, and FOM-MEQALAC systems provide an extremely useful technology base from which many future heavy-ion systems can be launched. Studies[24] on resonant-coupling mechanisms to improve longitudinal and azimuthal field stability in the presence of assembly tolerances and heavy beam loadings are timely and should be incorporated into future RFQ structures.

MISCELLANEOUS

Field stability and laboratory tolerance measurements are time consuming and sometimes difficult to accomplish. A useful computer code that models the RFQ geometry and shows good agreement with measured results (results from different Laboratories) was written and recently modified by Hutcheon.[25] The code can treat end tuning systems, VCRs, loop ports, misalignments, tube distortion and off-resonance effects, and can provide output in a concise and useful format.

Multiple rf drives to a cw RFQ structure may be necessary in the future for a number of reasons. A pulsed RFQ structure[26] is working without problems or difficulties as part of the H^- injector for the AGS at BNL. The structure is driven with about 120 kW via eight external coupling ports (two in each of four quadrants).

A feature that has been incorporated into most RFQ structures is the addition of VCRs between opposite vanes. The initial work[27] at LBL demonstrated the important consequences VCRs can have on maintaining field stability azimuthally between the four quadrants and shifting the dipole mode far enough in frequency from the operating mode to reduce interference effects on the longitudinal field distribution.

DISCUSSION

A good technological base has been established worldwide for the basic RFQ device--irrespective of the geometry selected for a particular application. Continued good communication and collaboration is a necessary ingredient for future progress that will be to everyone's benefit. On the basis of current information, the following items list the author's preference with respect to the design of a cw RFQ that must be reliable and have good performance characteristics.

- No manifold around RFQ structure
- Loop coupling directly to one of the quadrants
- Local compensation for penetration holes and perturbations
- Piston tuners in opposite quadrants
- No variable end tuners--just re-entrant end plugs
- End regions tuned to match structure frequency
- Vane coupling rings every λ length at a minimum
- Four-rod geometry if frequency <150 MHz
Four-vane geometry if frequency >150 MHz
- Resonant-coupling scheme at ends
- Penetration holes designed to minimize rf leakage
- Thermal properties considered in detail
- Pole tips from solid OFHC copper
- Welding and brazing employed as much as possible
- Removable vanes
- Cryogenic vacuum pumping
- Excellent diagnostics
- RF windows located to not see rf cavity region

ACKNOWLEDGMENTS

The author would like to thank H. Klein, D. Proch, H. Euteneur, S. Inagaki, N. Ueda, S. B. Hodge, J. Ungrin, G. E. McMichael, R. M. Hutcheon, M. A. Allen, D. A. Swenson, T. Smith, E. L. Kemp, J. M. Potter, D. D. Armstrong, W. D. Cornelius, T. P. Wangler, and R. H. Stokes for many interesting and illuminating discussions on cw aspects of accelerators over the past several years. A grateful thanks is conveyed to the many artisans, engineers, builders, and operators who have perfected the art of making reality out of many of our dreams.

REFERENCES

1. I. M. Kapchinskii and V. A. Teplyakov, "Linear Ion Accelerator with Spatially Homogeneous Strong Focusing," Prib. Tekh. Eksp. No. 2, 19 (1970).
2. Golosai et al., "Tests on the Initial Section of an Accelerator with Quadrupole HF Focusing," Atom. Energ. 39 (2), 123 (1975).
3. B. K. Gorshkov, "Startup of the URAL-15 Linear Proton Accelerator with Quadrupole rf Focusing," Sov. Phys. Tech. Phys. 22, No. 11, 1348 (1977).
4. R. H. Stokes, K. R. Crandall, J. E. Stovall, and D. A. Swenson, "RF Quadrupole Beam Dynamics," IEEE Trans. Nucl. Sci. 26 (3), 3469 (1979).
5. R. H. Stokes et al., "The Radio Frequency Quadrupole: General Properties and Specific Applications," Proc. XIth Int. Conf. on H. E. Accel., Geneva, Switzerland, July 1980, 399 (1980).
6. R. A. Jameson, "High-Intensity Deuteron Linear Accelerator (FMIT)," IEEE Trans. Nucl. Sci. 26 (3), 2986 (1979).
7. S. O. Schriber, "Electronuclear Fuel Production Using High Intensity Accelerators," Atomkernenergie 32, 49 (1978).
8. D. D. Armstrong, "Status Report on FMIT," Proc. 1984 Linac Conf., Gesellschaft für Schwerionenforschung report GSI-84-11, 505 (1984).
9. W. D. Cornelius, "CW Operation of the FMIT RFQ Accelerator," Proc. 8th Conf. on Appl. of Accel. in Res. and Ind., Nucl. Inst. Meth. Phys. Res. B 10/11, 859 (1985).
10. W. D. Cornelius, "CW Operation of the FMIT RFQ Accelerator," Proc. 1985 Part. Accel. Conf., Vancouver, Canada, May 1985, to be published.
11. J. McKeown, H. R. Schneider, and S. O. Schriber, "High Power Operation of Two Side-Coupled Standing Wave Linac Structures," Proc. 1972 Linac Conf., Los Alamos Scientific Laboratory report LA-5115, 233 (1972).
12. J. S. Fraser, J. McKeown, G. E. McMichael, and W. T. Diamond, "Operation of the Graded-β Electron Test Accelerator," Proc. 1976 Linac Conf., Atomic Energy of Canada Limited, Report No. AECL-5677, 166 (1976).
13. B. G. Chidley et al., "The Chalk River High Current Test Facility," Proc. of 1972 Linac Conf., Los Alamos Scientific Laboratory report LA-5115, 218 (1972).
14. J. Ungrin et al., "Low Energy 100% Duty Factor Alvarez Linacs at Chalk River," Proc. 1981 Linac Conf., Los Alamos National Laboratory report LA-9234-C, 12 (1981).
15. J. Ungrin et al., "Initial Operation of a 100% Duty Factor 3 MeV Alvarez Linac," IEEE Trans. Nucl. Sci. 28 (3), 3495 (1981).
16. R. M. Hutcheon et al., "Operation of a CW High Power RFQ Test Cavity: The CRNL 'Sparker,'" Proc. 1984 Linac Conf., Gesellschaft für Schwerionenforschung report GSI-84-11, 74 (1984).

17. B. G. Chidley and G. E. McMichael, "RFQ1 Design Parameters," ibid., 85 (1984).
18 J. Ungrin et al., "Construction of a High-Power RF Resonant Test Load," Proc. of 1981 Linac Conf., Los Alamos National Laboratory report LA-9234-C, 68 (1981).
19. R. M. Hutcheon, R. A. Vokes, T. Tran Ngoc, and J. C. Brown, "The RF Design of a 270 MHz, CW Four Vane RFQ," Proc. 1985 Part. Accel. Conf., Vancouver, Canada, May 1985, to be published.
20. R. M. Hutcheon and R. A. Vokes, "Penetration of RF Fields into Holes in Cavity Walls," ibid., to be published.
21. A Gerhard, H. Deitinghoff, H. Klein, and A. Schempp, "RF Sparking Experiments," Proc. XIth Int. Symp. on Disch. and Elect. Ins. in Vacuum, Berlin, GDR, September 1984, 37 (1984).
22. A. Schempp et al., "Zero-Mode-RFQ Development in Frankfurt," Proc. of 1984 Linac Conf., Gesellschaft für Schwerionenforschung Report No. GSI-84-11, 100 (1984).
23. P. Junior et al., "Investigations of Different RFQ Electrode Profiles for Easy Manufacture," ibid., 97 (1984).
24. A. Schempp, "Field Stabilization of RFQ Structures," ibid., 338 (1984).
25. R. M. Hutcheon, "An Equivalent Circuit Model of the General 3-Dimensional RFQ," IEEE Trans. Nucl. Sci. 30 (4), 3524 (1983).
26. H. Brown et al., "Design, Fabrication, and Testing of the BNL Radio Frequency Quadrupole Accelerator," Proc. of 1984 Linac Conf., Gesellschaft für Schwerionenforschung report GSI-84-11, 65 (1984).
27. D. Howard and H. Lancaster, "Vane Coupling Rings: A Simple Technique for Stabilizing a Four-Vane Radiofrequency Quadrupole Structure," IEEE Trans. Nucl. Sci. 30 (2) (1983), 1446.

RFQ WORK AT INS

N. Tokuda
Institute for Nuclear Study, University of Tokyo,
Tanashi, Tokyo 188, Japan

ABSTRACT

Research and development of heavy ion RFQ linacs at INS are described. We have constructed two four-vane RFQs, a test machine LITL and a real machine TALL, and a cold model of a split coaxial RFQ. The four-vane RFQs, driven with a single loop coupler at 100 MHz, accelerates ions with charge to mass ratio larger than 1/7. For very heavy ions, such as uranium, an improved structure of a split coaxial RFQ was proposed. This paper summarizes the operation of LITL, design, construction, and preliminary results of low power tests of TALL and the cold model of the split coaxial RFQ.

INTRODUCTION

RFQ linacs for heavy ions are developed at INS. For ions with charge to mass ratio, q/A, larger than 1/7, a couple of four-vane 100 MHz RFQ have been fabricated: a test machine LITL and a real machine TALL. The basic parameters of these linacs are summarized in Table I. LITL was constructed in 1982 to examine feasibility of driving the cavity with a single loop coupler, and to develop methods of vane setting and of rf contact between a vane and a tank. The acceleration test and the operation experience convince us that a four-vane RFQ is reliable as a preaccelerator for light heavy-ions. TALL is an extended version of LITL. This linac will work in a heavy-ion synchrotron complex. The beam is accelerated through the 7.3 m long linac up to 800 keV/u, and injected into a drift tube linac. The design principle of TALL is almost same as that of LITL, except for the longitudinal separation of the whole tank into four pieces. The cavity was assembled, and low-power tests on resonant frequency and field distribution are going on with good results.

For very heavy ions, such as uranium, an improved structure of a split coaxial RFQ was proposed. At this structure, mechanical stability of electrodes is improved by supporting them to the tank by stems at several points longitudinally, and modulated vanes are mounted on the electrodes to obtain a well formulated electric field. A cold model with unmodulated vanes was fabricated to study rf characteristics of this structure, and satisfactory field distribution has been obtained.

This paper describes experience in high-power operation of LITL, mechanical design and preliminary results of low-power tests of TALL and the cold model of the split coaxial RFQ. More detailed descriptions on TALL and the split coaxial RFQ are presented elsewhere.[1,2]

TABLE I. Basic parameters of LITL and TALL.

Parameters	LITL	TALL	Units
Charge to mass ratio of ions, q/A	1/7	1/7	
Frequency	100	100	MHz
Input energy	5	8	keV/u
Output energy	138	800	keV/u
Vane length, L	122	725	cm
Cavity diameter, D	56	58	cm
Characteristic bore radius, r_o	0.41	0.54	cm
Minimum bore radius, a_{min}	0.25	0.29	cm
Intervane voltage, V	62	81	kV
Maximum surface field	205	205	kV/cm
Kilpatrick factor	1.8	1.8	
Wall loss of rf power (60% Q)	22	180	kW
Output current	7	7	emA

LITL

Summary of operation

The first acceleration through LITL was performed in July, 1982. Since then, beam experiments have been done on transmission efficiency, output beam emittances in transverse and longitudinal phase spaces, *etc.* The experimental results agree well with the PARMTEQ simulation.[3]

During these three years, LITL worked for 300 h with beam, and 100 h without beam. Meanwhile, we have no severe problem arisen from secular changes in the Q-value and the field unbalance. In 1982, the Q-value measured 10600, which is 60% of the SUPERFISH value assuming an ideal cavity. This Q-value stays constant, though the cavity has been exposed to the air 20 or 30 times. The field unbalance among the four quadrants was tuned initially within ±2% in July, 1982. This unbalance went larger with time: ±7% at measurement in January, 1985. The field balance recovered to the initial value easily with side tuners, aluminum-alloy plungers of 85 mm in diameter.

Sparking test

A sparking test was performed by exciting the LITL cavity around 2 Kilpatrick. The designed intervane voltage is 62 kV, at which the maximum surface electric field is 205 kV/cm, corresponding to 1.8 Kilpatrick. The required rf power for the cw operation is 22 kW. In the sparking test, the stability of the cavity was examined for various values of intervane voltage and a duty factor. The result is summarized in Table II. The cavity was stable at 1.9 Kilpatrick and a duty factor of 40% (Run No. 5). At the cw operation at 1.84 Kilpatrick (Run No. 6), no sparking took place for beginning two minutes; thenceforth, sparking lasts intermittently.

TABLE II. Sparking test with the LITL cavity.

Run No.	Intervane voltage (kV)	Kilpatrick	Pulse width Interval (ms/ms)	Duty factor (%)	Note
1	66.6	1.94	1/25	4	stable
2	60.6	1.76	10/50	20	stable
3	63.1	1.84	10/50	20	stable
4	65.4	1.91	20/50	40	stable
5	70.3	2.05	1/10	10	sometimes spark
6	63.1	1.84	cw	100	stable for 2 min

Though no further sparking test has been performed to avoid vane tip's damage, we consider that cw operation at 1.8 Kilpatrick could be attained with sufficient aging and cooling of the cavity.

Variation of resonant frequency and temperature

Variation of resonant frequency and tank temperature during cw operation at about 1/3 of the full power were measured. Two thermometers were attached on the end wall and the tank's side wall as shown in Fig. 1. Time variations of the resonant frequency and the tank temperatures are shown in Fig. 2. At the beginning stage, operation at 6.9 kW, the resonant frequency decreased rapidly by 250 kHz in 10 min. The temperature at the tank's side wall increased by 15°C in 40 min, and stayed constant. On the other hand, the temperature at the end wall, which is not cooled with water, increases gradually with a constant slope, and reaches 58°C at the power-off. When the rf power was increased to 7.6 kW at 115 min, the resonant frequency decreased by 50 kHz in response to it. The vacuum pressure at the pump head, 7×10^{-7} Torr initially, increased to

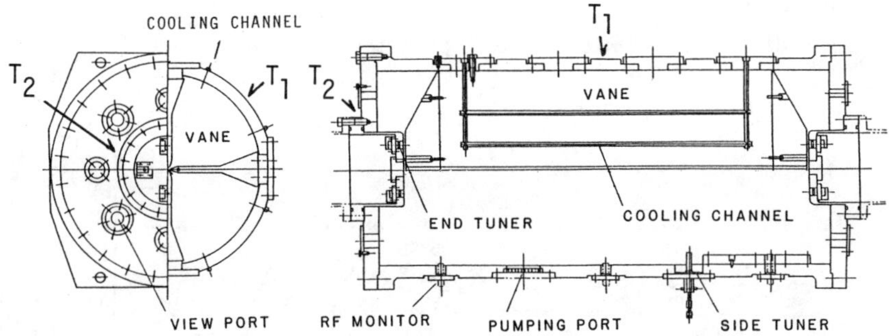

Fig. 1. Cross sectional view of LITL. Two cooling channels run through each vane, and eight ones on the tank's side wall. An oil-filled thermometer is attached on the tank's side wall, T_1, and a thermocouple on the end wall, T_2, to measure temperature increase at high-power operation.

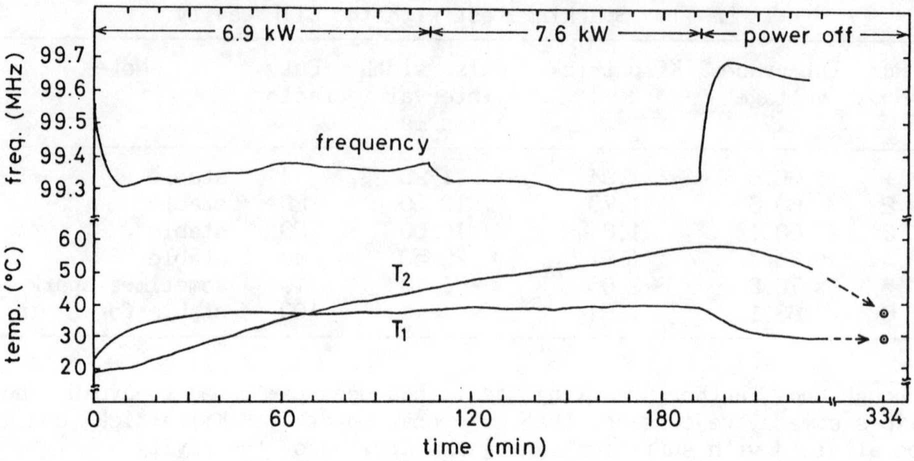

Fig. 2. Time variation of the resonant frequency, f, the temperature on the tank's side wall, T_1, and the one on the end wall, T_2, during high-power operation of LITL.

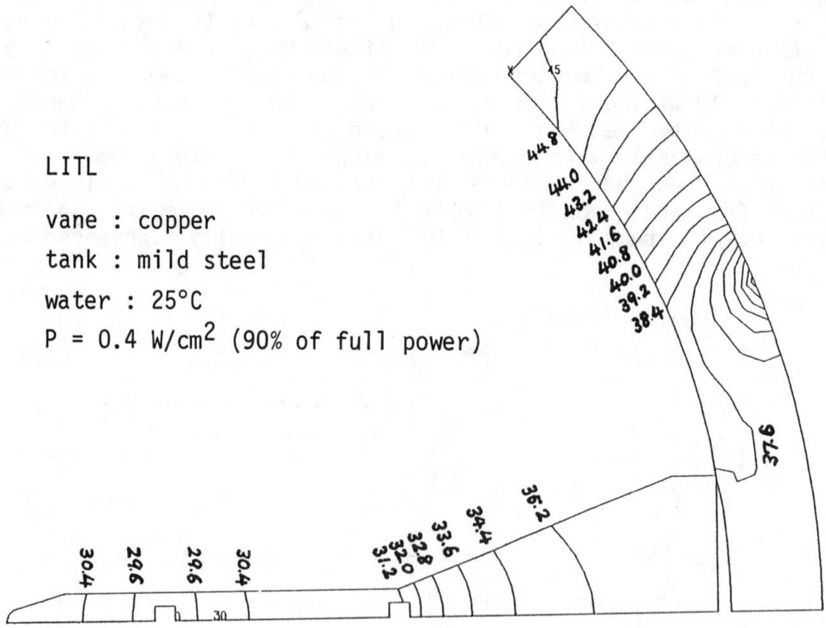

Fig. 3. Numerically calculated temperature distribution in the vane and the tank of LITL.

1.2×10^{-6} Torr at the 6.9 kW operation, and to 2×10^{-6} Torr at 7.6 kW. The pressure recovered to 9×10^{-7} Torr two hours after the power-off.

These data on temperature imply that the vanes are heated appreciably. The expansion of vanes leads to increase of capacitance between vanes, and, in result, to decrease of the resonant frequency. This frequency shift is roughly estimated using an equation,[4]

$$\ln\left(\frac{ka}{2}\right) + \frac{2}{k^2R^2} + \gamma = 0 \ , \qquad (1)$$

where

k : wave number = 0.02096 cm^{-1},
a : bore radius = 0.41 cm,
R : tank radius = 28 cm,
γ : Euler's constant = 0.5772... .

Assuming the tank is heated uniformly by ΔT_t, and the vanes by ΔT_v, we have an equation for frequency shift,

$$\frac{\Delta f}{f} = \left(\frac{4}{k^2R^2} - 1\right)^{-1}\left[\left(\frac{R}{a} - \frac{4}{k^2R^2}\right)\alpha_t \Delta T_t - \left(\frac{R}{a} - 1\right)\alpha_v \Delta T_v\right] , \qquad (2)$$

where α_t is the expansion coefficient of the tank material, and α_v is that of the vane material: $\alpha_t = 1.2 \times 10^{-5}\ ^\circ\mathrm{C}^{-1}$ (mild steel), and $\alpha_v = 1.7 \times 10^{-5}\ ^\circ\mathrm{C}^{-1}$ (copper). With experimental values for ΔT_t and Δf, the temperature increase of the vanes is estimated at about 30°C.

This estimated temperature increase of the vanes contradicts a temperature distribution given by a computer calculation. Figure 3 shows its result for 25°C cooling water and power dissipation of 0.4 W/cm^2, or 19.5 kW in total, about 90% of the full power. Even at this high power operation, the temperature increase of the vane is less than 10°C. We could guess that the actual vane cooling is inefficient, but cannot conclude so, because we have not measured the temperature distribution on the vanes. The measurement of temperature is now planned.

TALL

Design features of the cavity

The TALL cavity is 58 cm in diameter and 730 cm in length. The tank length, L, is 2.4 times as long as the wavelength in free space, λ. The cavity, both of the tank and the vanes, is longitudinally divided into four pieces. This separation is effectual in achieving not only easier machining and assemblage of the tank and the vanes but better field distribution. Vane's warps and positioning errors affect vane voltage by their effect on the local capacitance error between vanes;

$$\frac{\Delta V}{V_0} \propto \left(\frac{L}{\lambda}\right)^2 \frac{\Delta C}{C_0} \ , \qquad (3)$$

where ΔV is the local error of voltage, and ΔC is that of capacitance.[5] The factor $(L/\lambda)^2$ of 5.8 is so large at TALL, 0.7 at LITL, it is necessary to reduce the capacitance error by fabricating the vanes and setting them precisely. For this purpose, the vanes are divided longitudinally into short pieces and set to each of the the four tanks. The resulting error of the vane setting is averaged over the whole cavity, and voltage deviation is expected to be smaller than that of a cavity without the longitudinal separation, because L in Eq. 3 is effectively replaced with $L/4$.

Considerations on longitudinal vane gap

The vanes are set with a 0.2 mm gap between longitudinally adjacent ones. The vane's end is chosen at a crest or a trough of modulation ($kz = 0$ or π). No voltage jump is expected to occur at the gap, because current does not flow longitudinally at these vane's end. The effects of the gap on the field distribution, the resonant frequency, and the Q-value were examined with a 300 MHz RFQ cavity where the vanes are installed with longitudinal gaps. The experimental results show that these effects are insignificant, so far as the longitudinal field distribution is tuned to be flat.[6]

The enhancement of surface field at the gap was roughly estimated the SUPERFISH code.[7] For vanes with a 0.2 mm gap and circularly shaped corners of 0.3 mm radiuses, the enhancement was 1.24. This leads to the maximum surface field of 1.54 V/r_o, as the maximum surface field at the crest and the trough of a vane without a gap is 1.24 V/r_o. This enhanced maximum surface field is nearly equal to that at quadrupole symmetry ($kz = \pi/2$), 1.56 V/r_o. Therefore, we consider the field enhancement at the gap is not serious.

Alignment of vanes

TALL is now installed with aluminum vanes without cooling channels. The vanes are mounted to the tank in the same manner as in LITL. The rf contactors are stainless steal C-rings, 3.2 mm in diameter and 0.35 mm in thickness, plated with silver to a thickness of 50 μm. These vanes will be replaced with the ones with cooling channels for high-power operation, which are now under design.

For each of the four tanks, vanes are set with an accuracy better than ± 50 μm. The positions of the vanes are measured from two fiducial lines on the square frange of the tank: the bottom rim and one of the side rims. Then the tanks are set on a bed, of which the surfaces are leveled within ± 20 μm. The resulting alignment error of the beam axis over the whole length of 7.3 m is 0.2 mm, and the step between longitudinally adjacent vanes is smaller than 0.1 mm. According to a computer simulation, this alignment error has little effect on beam transmission, as shown in Fig. 4.

Field tuning

Resonant frequencies of various modes are measured by shorting

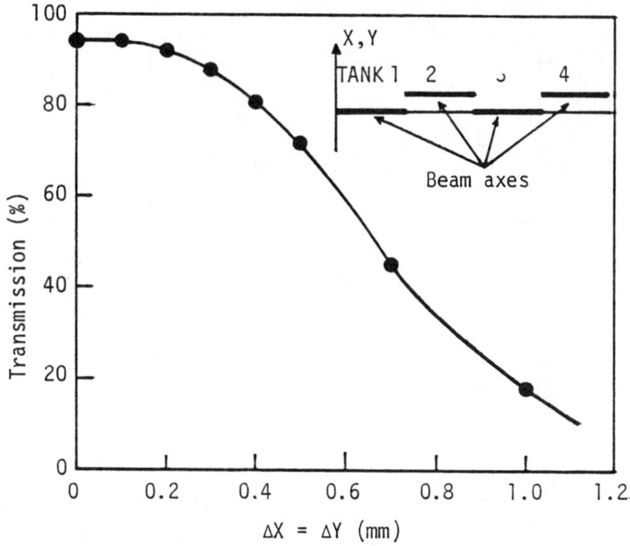

Fig. 4. PARMTEQ result of transmission efficiency as a function of displacement of the beam axis. The displacement within 0.1 mm has little effect on the transmission.

the vanes to the cavity end wall. Figure 5 shows the resulting dispersion curves for the TE11 and the TE21 modes. The obtained cutoff frequency of the TE11 mode is 97.6 MHz, and that of the TE21 mode is 101.1 MHz. These values differ slightly from SUPERFISH values, 98.3 MHz and 100.6 MHz, respectively. These differences are considered to come from the effect of vane modulation.

Longitudinal voltage distribution was, at first, tuned with inductive tuners attached at the ends of the vanes. The cross sectional area for return of magnetic flux at the cavity end wall was varied and set at about 1/4 of the area of one quadrant. Then, the longitudinal field uniformity and the azimuthal field balance are improved with the end capacitive tuners and 24 inductive side tuners. Figure 6 shows magnetic field distribution before and after the preliminary tuning. Tentative uniformities are $\pm 4.9\%$ longitudinally, and $\pm 5.7\%$ azimuthally. These values will be improved by further fine tuning. The resulting resonant frequency of the TE210 mode is 101.78 MHz, which is 0.93 MHz lower than that of the nearest TE11 mode.

Research schedule

After finer tuning of field distributions, an acceleration test will be performed in coming summer and fall, 1985, with light ions, such as protons and H_2^+, by exciting the cavity with the 25 kW rf power source for LITL. A high power test is scheduled after the mounting of the vanes with cooling channels, which is under preparation, and purchase of a high power rf source.

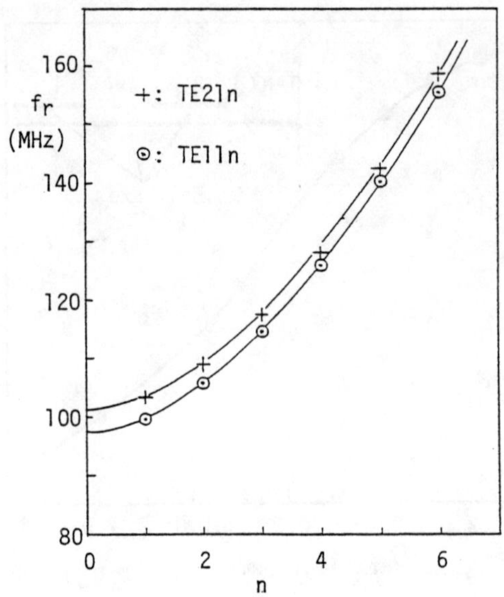

Fig. 5. Dispersion curves of the TALL cavity, obtained by shorting the vanes to the end wall. The cutoff frequency is 97.6 MHz for the TE11 mode, and 101.1 MHz for the TE21 mode.

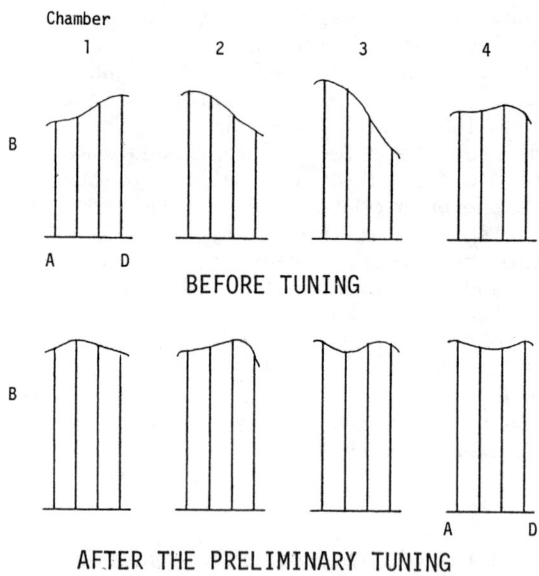

Fig. 6. Magnetic field distribution in the TALL cavity, before and after the preliminary tuning. The resulting field uniformity is ±5.7% azimuthally, and ±4.9% longitudinally.

SPLIT COAXIAl RFQ

Design and fabrication

The features of our split coaxial RFQ are: (1) modulated vanes, similar to those in a four-vane RFQ, are used as electrodes, and generate a well formulated electric field; (2) electrodes are supported by stems to the tank to attain good mechanical stability of the vanes.[8] The configuration of a vertical and a horizontal vane is schematically shown in Fig. 7. The voltage difference between the vertical and the horizontal vane is approximately constant over the whole length.

The fabricated cold model of the split coaxial RFQ is a 1/4-scale model of a 12.5 MHz machine, which will accept U^{4+} ions. The cavity, made of brass, is 2 m in length and 0.4 m in diameter. Unmodulated vanes are installed with a bore radius of 5 mm. The tank is divided into four pieces for the convenience of mounting the electrodes. As shown in Fig. 8, each vane is attached to a diamond-shaped beam, which is supported by a stem bolted to a tank at its frange. To ensure the relative positioning of the four vanes, opposite vanes are connected with coupling rings at four places longitudinally (eight places in total for vertical and horizontal couples). The vanes, the beams, the stems, and the coupling rings are assembled in a body, and the vanes are set with an accuracy better than ±0.05 mm. Then, the assembly is installed in the tank.

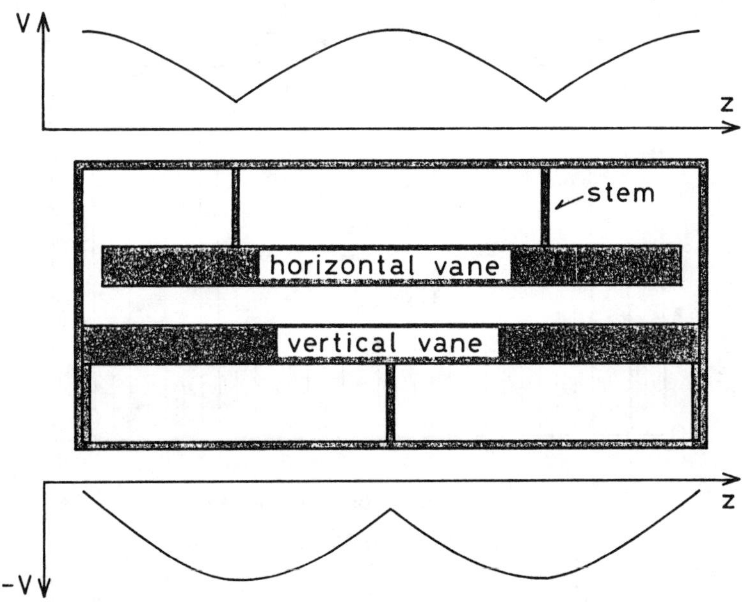

Fig. 7. Configuration of electrodes and qualitative longitudinal voltage distribution in the split coaxial RFQ.

Fig. 8. Assemblage of the cold model of the split coaxial RFQ.

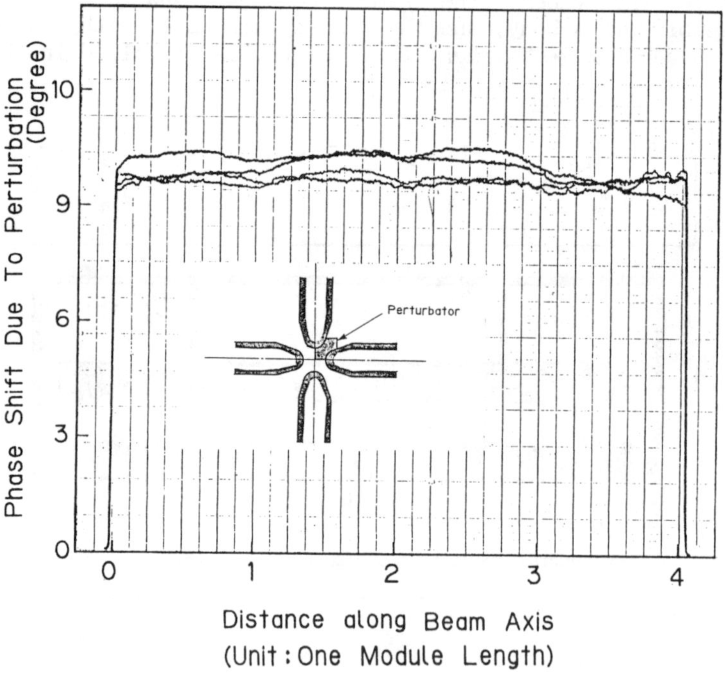

Fig. 9. Longitudinal field distributions in the four quadrants. The flatness of the longitudinal field distribution is better than ±2%, and the unbalance among them is smaller than ±2.5%.

Field distribution

A low-power test on field distribution provides satisfactory results both in longitudinal field distribution and azimuthal field balance.

Figure 9 shows longitudinal field distributions in the four quadrants. A teflon perturbator was moved longitudinally touching two adjacent vanes. The flatness of the longitudinal field distribution is better than ±2%, and, the unbalance among the quadrants is less than ±2.5%. The resonant frequency measures 37.12 MHz, which is 26% lower than the designed value. This discrepancy is mainly attributed to the coupling rings, which were not considered at the design stage of the cavity, but attached afterward to improve the the accuracy of vane positioning.

Future study

A good result have been obtained in field distribution. The resonant frequency is, however, still a problem to be solved. An equivalent circuit describing the split coaxial cavity is now under formulation. After this problem is solved, the cavity will be mounted with modulated vanes for an acceleration test.

ACKNOWLEDGMENTS

The author would like to thank S. Arai, T. Nakanishi, N. Ueda, and S. Yamada for their supplying experimental data, helpful discussions, and comments on the manuscript.

REFERENCES

1. N. Ueda *et al.*, Proc. 1985 Particle Accelerator Conference, Vancouver, Canada, May, 1985, to be published in October 1985 IEEE Trans. Nucl. Sci.
2. S. Arai *et al.*, *ibid.*
3. N. Ueda *et al.*, IEEE Trans. Nucl. Sci., Vol. NS-30, No. 4, 2975 (1983).
4. H. Lancaster and K. J. Kim, LBL-Medical Accelerator Note 13 (1981).
5. F. O. Purser *et al.*, IEEE Trans. Nucl. Sci., Vol. NS-30, No. 4, 3582 (1983).
6. N. Ueda *et al.*, Proc. 5th Symp. Accel. Sci. Tech., Tsukuba, Japan, 86 (1984).
7. T. Nakanishi *et al.*, *ibid.* 89.
8. S. Arai, GSI-83-11 (1983).

NONINTERCEPTIVE BEAM DIAGNOSTICS*

D. D. Chamberlin, AT-4, MS H821
Los Alamos National Laboratory
Los Alamos, NM 87545

ABSTRACT

The need for accurate real-time diagnostics is critical for high-power particle beams. This paper describes the present level of development of noninterceptive devices for these beams. Discussion will center on diagnostic measurements as they occur along the beamline, from ion-source to the input into an RFQ, and on measurements of the RFQ output, using the cw beam at Los Alamos as an example.

INTRODUCTION

The new generation of high-power particle accelerators will require noninterceptive diagnostic techniques and sensors to handle their special requirements. An intrusive sensor would be damaged or, more likely, destroyed, and appreciable radiation would be produced by the interception of some fraction of the beam. Yet the need for accurate real-time diagnostics is more critical because the beam power must be well monitored and controlled. The melting power of these beams is impressive--even for regions considered to be the distant edges ("halo") of the beam.[1]

The diagnostic requirements for the ion injector are quite demanding and are representative of the overall accelerator needs. This paper describes the present development of new diagnostic devices for high-power beams limited to an ion beam of less than 5 MeV, currents of greater than 20 mA and pulse lengths of 1 ms or more. The sensors and techniques fulfill the several purposes of assisting in machine commissioning and operation, verifying beam-dynamics calculations, and minimizing beam losses. The devices are grouped by the parameter to be measured or derived, including current, centroid location, beam profile and distribution, emittance (transverse and longitudinal), and energy.

The physical environment in the beam region before the RFQ is very different from that after the RFQ (or other first rf structure). Prior to the RFQ, the beam is a low energy, high-perveance dc beam (during a macropulse) with a strong susceptibility to emittance growth (a "soft" beam). There may be small, high-frequency plasma oscillations but no long term time structure. The beam interacts strongly with the relatively dense residual gas in a beam pipe that has ample space for sensor installation. After the RFQ, the beam has greater energy, is longitudinally bunched with full-scale megahertz time structure, and is much less

*Work supported by the US Dept. of Energy.

susceptible to emittance growth (a "stiffer" beam). The electromagnetic interactions with the background gas are weaker and the proximity of the succeeding linac may imply severe limitations on sensor placement.

This paper will treat diagnostic measurements as they occur along a beamline, from the ion-source to the input of the RFQ, and measurements at the RFQ output, for the cw beam at Los Alamos. Although references to work at several laboratories are included, this is not an exhaustive review of the literature. A previous review of nondestructive diagnostics by Fraser, and a recent summary of technology status is available.[2]

DC ION CURRENT

The basic measurement for an ion source is detecting the extracted beam as shown by charge collection in a Faraday cup or by visual observation of the background gas fluorescing in locations excited by the beam particles. A Faraday cup and a viewing porthole should be positioned quite near the source. Direct electrical measurements of impinging charged particles are unreliable because of the extremely voltage-sensitive flow of ions and electrons over the length of the low-energy transport line. The FMIT prototype accelerator[3] at Los Alamos has used a single-loop ferrite toroid, a transductor, to pick up the second-harmonic modulation of the beam's magnetic field.[4] Daily usage of this transductor has been generally satisfactory except for the frequent need to reset the absolute zero current level. A next-generation transductor should have two toroidal elements with an attendant self-zeroing feature. Such a configuration would be an adaptation of the many-element current monitors developed at the CERN laboratory by Unser.[5]

BEAM PROFILES

The traditional measurement of beam current versus relative transverse position performed with the technique of a wire scanner passing through the beam is usually not feasible for high-power beams because of damage or melting of the wire. A beam-profile may be obtained from an optical sensor (for example, a TV camera) responding to the light produced by beam interactions with residual gas. These interactions produce visible light primarily through the excitation and immediate (life-time of approximately 60 ns) decay fluorescence from levels in the N_2^+ ion with most of the intensity near 400 nm.[2,6] The spectral content and the intensity of this excitation light are dependent on residual gas species and pressure. With N_2 background gas at 10^{-6} Torr and a 20-mA H_2^+ beam at 75 keV, adequate light is produced for a silicon intensified target (SIT) tube in a TV camera. The correlation of calculated light intensity with beam particle intensity has not been precisely demonstrated, but indications are that the agreement is within a few percent.[7] The sample beam profile shown in Fig. 1 of Ref. 8 obtained from a digitized TV line-scan demonstrates a four-to-one signal-to-background ratio without any special optimization. These profiles may be transferred rapidly over data links for recording or display, although the raw TV picture is greatly valued by operators for beam tuning. The beam profile displays can be updated at a several-per-second rate, thus providing the operator with a real-time diagnostic For low light situations, the updating rate depends on the light intensity and the required integration period of the light-sensitive elements for a pre-determined signal-to-noise ratio. Photodiode sensors can be cooled with thermoelectric pads to reduce their dark current. Microchannel plate (MCP) image intensifiers have been used to multiply the incident photon (or, interchangeably, the electron) flux.[9] Filters might be considered for selection or rejection of particular spectral lines. The literature does not have a readily usable rule of thumb for estimating light levels and signals in varying situations. An early measurement of the optical brightness provided by a proton beam at Los Alamos gave a radiance value of 91 mW/(sr-m^2), with an uncertainty of a factor of two, for a current density of 100 mA/cm^2 at 100 KeV, through 10^{-5} Torr of residual air (mostly N_2). This value may be scaled to other beam operating conditions; linearly with current, linearly with pressure, inversely with beam area and inversely with energy to the 0.8 power. The cross section for atomic excitation as a function of ion energy may be found in Ref. 10, or can be scaled from electron cross sections at lower energies, corresponding to equal electron and proton velocity.

The physical arrangement for obtaining optical profiles can be as simple as a TV camera viewing through a porthole in the beam pipe, or a complex arrangement of portholes and mirrors that images four views simultaneously onto the camera lens. These arrangements and a further step involving a four-legged spatially coherent fiber-optic bundle transferring beam images to two linear photodiode arrays have been reported in the proceedings of accelerator conferences in recent years.[2,6,8,9] The visible background from light reflections inside the beam pipe has been a problem and wavelength filters are being added. Related schemes have been reported for detection of synchroton radiation in large circular accelerators.[11] The whole field of optical sensing with fibers, photodiodes, and intensifiers is moving very rapidly, and order-of-magnitude enhancements in capability are reported every few months.[12]

SPATIAL DISTRIBUTIONS

A tomographic reconstruction of the particular density distribution that projects into the measured beam intensity profiles can be obtained from four measured transverse beam profiles by using a maximum entropy (smoothest distribution) criterion.[13] The determined density distribution is not a unique solution of the input profiles, even for the multitude of profiles measured in medical tomographic scans and fit using an Abel transformation or fast Fourier transform techniques. The maximum-entropy criterion leads to a distribution solution from four views in about four iterations of less than a second each in a CAMAC-based microcomputer. The algorithm has been well checked and verified for accuracy, stability, convergence, and noise rejection.[14] This maximum entropy criterion has been used for reconstructions in applications at Los Alamos, at the Swiss Nuclear Institute in Switzerland, and at Culham Laboratory in England.[15] Figure 3 of Ref. 9 shows a typical solution as an isometric representation of the cw beam from the 2-MeV RFQ at Los Alamos.

TRANSVERSE EMITTANCE

Analogous to the tomographic reconstructions of spatial density distributions are the reconstructions of the transverse emittance density distributions. The criteria for such calculations were stated by Metzger,[16] and the maximum-entropy algorithm described above provides a framework for this type of emittance reconstruction. For a spatial distribution reconstruction the code connects the four different profiles from different angles at one location by rotation matrices appropriate to the angles of the views.[8,13] For the emittance reconstruction, the three profiles from different longitudinal positions (all at the same angle) are connected using the appropriate beam-transport matrices. These transport matrices are taken from the dynamics code TRACE and incorporate all the features of that code, including space charge. In Ref. 9, Figure 2 shows the three distribution contours that describe the transverse emittance for that beam.

CENTROID POSITION

Position measurements locate the current or charge centroid of the beam within the beam pipe. This information is available from the beam profiles obtained by the optical sensors at their updating rate. After the RFQ, faster response is available from electromagnetic pickups. The difficulties with measuring dc beams and ion flows in the beam pipe preclude electromagnetic measures before the RFQ. The full-scale modulation of the current after an RFQ is detectable with magnetic loop pickups, with capacitive probes, with strip-line technology and with resonant cavity excitation.[17] A particular version of the magnetic-loop pickup is shown in Fig. 6 of Ref. 18. The dominant restriction was to avoid protruding into the cavity of the beam pipe where stray beam may destroy the sensor wire or the vacuum seal. The four-lobed structure fits within or between the four iron poles of a quadrupole magnet, even in the bore of a drift tube in an Alvarez linac. The first use of this structure[1] has shown that upstream-facing surfaces on which stray beam can impinge are quickly heated to the failure point of some component, such as a nearby O-ring.

BEAM CURRENT

Measuring the absolute level of the beam current after the RFQ is accomplished with a transformer coupled to the beam. For a pulsed beam, the rise and fall of magnetic field associated with the beginning and end of the pulse structure provides the observed signal. For a cw beam, the only time variation of the beam current is that of the rf microstructure, which calls for employing high-frequency techniques and an air-core transformer. A current transformer responsive to 80-MHz time structure has been specially constructed by Pearson Electronics and is in use at Los Alamos.[18] These diagnostic measurement transformers must be calibrated by calorimetric means whenever possible. A further possibility for the measurement arises from creating a macropulse time structure by deleting beam for a microsecond at a few-hertz rate.[19] This artificial pulse structure can be matched to transformer capabilities.

HALO OF THE BEAM

The fringes of the beam region may be probed with more traditional interceptive techniques if the thermal loading and required cooling are accounted for in the design and construction of the sensor. A nominal description of an ion beam profile is obtained by fitting a Gaussian envelope to the central core of the beam. The beam halo comprises the edges of the beam outside the two sigma radius of that Gaussian envelope where there is more beam, by a factor of a few, than the Gaussian description would imply. The interceptive measurements are absolute if taken calorimetrically, but are of questionable validity and repeatability if taken electrically. A difficulty with electrical measurements is the uncertainties in the production of secondary electrons and in the confinement or repulsion of those electrons by bias voltages applied to guard rings near the sensor.

The cw program at Los Alamos is working toward an overlap between optical and interceptive techniques, wherein the halo sensor and the light sensor may both be responsive to the same region outside the core.

ENERGY

The kinetic energy of particle beams is established by the accelerating structures through which they have passed. A measurement of beam energy to high resolution is accomplished traditionally by bending the particle trajectories through a chosen angle in a magnetic field. Examples include the classical 90° analyzing magnets which maintain an energy with regulating slits, and the several geometric variations of a magnetic spectrometer. A noninterceptive scheme for deriving the average beam energy (and other longitudinal parameters) is to measure the flight time of a particular beam bunch over a known distance. Timing resolution of both capacitive pickups and electronics modules is in the ten-to hundred-picosecond range and steadily getting better. The fast sensor may be any of the electromagnetic techniques mentioned earlier, with the added consideration that smaller size implies greater speed of response. Fast electromagnetic sensors specially built by NTG[20] with a clear bore of eighteen cm. are being used on the cw beam line at Los Alamos. Discriminators coupled with fast, commercial, time pick-offs, and time-to-digital converters are providing energy updates for the operators within one second and with a resolution of better than one per cent.

LONGITUDINAL PARAMETERS

The fast sensors of the energy measurement offer further information on the longitudinal structure of an individual beam bunch.[21] The shape of the electric signal may be thought of as having a contribution from the characteristics of the sensor itself and a contribution from the charge distribution that excites it. The contribution of the charge distribution may be separated (deconvolved) from the composite using Fourier analysis.[22] The present analytic work on this technique at Los Alamos has progressed to the limit of the relatively large physical size and attendant slow response of the sensor.

PHOTODETACHMENT

Of special interest to H^- ion injectors is a newly demonstrated technique[23] for gently detaching the electron and observing the now neutral H° atom after it drifts free of the region of applied electromagnetic fields. The neutralization is by photodetachment of the electron by a 1.06-μm photon from a Nd:Yag laser. There is a tremendous selectivity in this technique in the time and space focusing capability of the laser beam. Figure 1 of Ref. 23 illustrates that selectivity as a 32-ps slice in the longitudinal plane and as several or tens of microns spatial focus. As a tool for the future, note that this technique can explore the internal conditions of an RFQ or an Alvarez linac.

SUMMARY

Noninterceptive diagnostic techniques have been implemented in early versions which provide much of the needed data on conditions and performance of a high-power beam. The optical sensing of beam-induced fluorescence is a key element in a diagnostic system for providing spatial distribution and transverse emittance information. Specific characteristics of individual accelerators and their beams offer tradeoffs among the techniques. The applications of these noninterceptive sensors are expected to show dramatic improvements as the field matures.

These improvements should cover speed of response, physical size of the installation, dynamic range over beam current, computer interface of data, cost, system complexity and resolution of the data.

REFERENCES

1. W. D. Cornelius, "CW Operation of the FMIT RFQ Accelerator," Proc. 1985 Particle Accelerator Conf., IEEE Trans. Nucl. Sci. 32 (5), (1985), in press.
2. J. S. Fraser, "Developments in Non-Destructive Beam Diagnostics," IEEE Trans. Nucl. Sci. 28 (3), 2137 (1981) and D. D.

Chamberlin, "Summary from Working Group on Noninterceptive Diagnostics," Los Alamos National Laboratory document LA-UR-85-2942 (1985).

3. D. D. Armstrong, "The FMIT Accelerator," IEEE Trans. Nucl. Sci. 30 (4), 2965 (1983).
4. A. T. Brousseau and D. D. Chamberlin, "FMIT Direct-Current Beam Monitor," IEEE Trans. Nucl. Sci. 28 (3), 2299 (1981).
5. K. Unser, "A Toroidal DC Beam Current Transformer with High Resolution," IEEE Trans. Nucl. Sci. 28 (3) 2344 (1981).
6. D. D. Chamberlin, G. N. Minerbo, L. E. Teel, and J. D. Gilpatrick, "Noninterceptive Transverse Beam Diagnostics," IEEE Trans. Nucl. Sci. 28 (3), 2347 (1981).
7. J. F. Bonnal, et al., "Optical Diagnostics on High-Power Neutral Beams," Physics Letters, 69A, 116, (1978).
8. D. D. Chamberlin, G. N. Minerbo, and C. T. Mottershead, "Noninterceptive Transverse Beam Measurements," Proc. 1981 Linac Conf., Los Alamos National Laboratory report LA-9234-C, 159 (1981).
9. D. D. Chamberlin, J. S. Hollabaugh and C. J. Stump, Jr., "Imagescope to Photodiode Beam-Profile Imaging System," IEEE Trans. Nucl. Sci. 30 (4), 2201 (1983).
10. W. W. DeLuca, "Beam Detection Using Residual Gas Ionization," IEEE Trans. Nucl. Sci., 26 (3), 813 (1969).
11. J. Bosser, et al., "Beam Intensity, Position and Profile Measurements for CERN SPS Collider," Proc. 1985 Particle Accelerator Conf., IEEE Trans. Nucl. Sci. 32 (5), (1985), in press.
12. R. A. Wich, "VIM: Grabbing Images at Very Low Light Levels," Photonics Spectra, 19 (5), 133, (1985).
13. C. T. Mottershead, "Maximum Entropy Beam Diagnostic Tomography," Proc. 1985 Particle Accelerator Conf., IEEE Trans. Nucl. Sci. 32 (5), (1985), in press.
14. G. N. Minerbo, "MENT, a Maximum Entropy Algorithm for Reconstructing a Source for Projection Data," Computer Grapics Image Processing 10, 488 (1979).
15. G. A. Cottrell, "Tomography of Neutral Beams," Rev. Sci. Instrum. 55 (9) 1401 (1984).
16. C. Metzger, "Measurement of Beam Emittance and Beam Centering in the 800-MeV Measurement Line of the PS Booster,"/CERN/SI/Int. DL/69-10.
17. J. McKeown, "Beam Position Monitor Using a Single Cavity," IEEE Trans. Nucl. Sci., NS-26 (3), 3423 (1979).
18. J. D. Gilpatrick and D. D. Chamberlin, "FMIT Diagnostic Instrumentation," Proc. 1985 Particle Accelerator Conf., IEEE Trans. Nucl. Sci. 32 (5), (1985), in press.
19. H. Herminghaus, et al., "The Design of a Cascaded 800 MeV Normal Conducting C.W. Race Track Microtron," Nucl. Instrum. Meth. 138, 1 (1976).
20. Nukleartechnic GMbH, 6460 Geimhausen 2 Hailer, West Germany.
21. P. Strehl, "A New Method for Longitudinal Emittance Measurements," IEEE Trans. Nucl. Sci. NS-30 (4), 2198 (1983).

22. J. D. Gilpatrick and C. A. Watson, "Reconstruction of Longitudinal Beam Profiles from Nondestructive Electromagnetic Probes," Proc. 1985 Particle Accelerator Conf., IEEE Trans. Nucl. Sci. 32 (5), (1985), in press.
23. W. B. Cottingame, et al., "Noninterceptive Techniques for the Measurement of Longitudinal Parameters for Intense H^- Beams," Proc. 1985 Particle Accelerator Conf., IEEE Trans. Nucl. Sci. 32 (5), (1985), in press.

HIGH-CURRENT BEAM TRANSPORT AND CHARGE-NEUTRALIZATION EFFECTS

M. Reiser
Laboratory for Plasma and Fusion Energy Studies, University of Maryland, College Park, Maryland 20742

ABSTRACT

The present status of theoretical and experimental studies on high-intensity beam transport is reviewed. Current and brightness limitations, emittance growth, nonlinear and charge-neutralization effects and highlights of ongoing experiments at the University of Maryland, Berkeley and GSI Darmstadt are discussed.

INTRODUCTION

High-current, high-brightness ion beams are presently of great interest for several applications such as heavy ion inertial fusion (HIF), spallation neutron sources, and injector linacs for future high-energy physics machines. The focusing, or transport, of such beams is dominated by space charge, rather than emittance, i.e. one operates near the space-charge limit where the forces due to the collective internal fields are comparable to the external focusing forces. Space-charge effects are most pronounced at low energy, and the only machines capable of accelerating low-energy beams near the space-charge limit are rf linacs or induction linacs. When the beam pulse width is sufficiently long, charge neutralization via collision with residual gas molecules may play an important role in the low-energy transport system.

In a broad sense, beam transport involves the transverse and longitudinal focusing of the particles through the accelerator to the target. The beam is produced in a high current, high-brightness source and then transported from the source to the accelerator. In the low-energy transport section the beam has to be pre-accelerated, bunched (with a chopper/buncher system), and matched into the linac. In present-day ion accelerators this function is accomplished with the Radio-Frequency-Quadrupole Linac (RFQ) serving as injector for an rf linac.

In the rf linac or induction linac acceleration, transverse, and longitudinal focusing actions occur at periodic time intervals. Thus, in the frame moving with the particles, the linac is a periodic beam transport system. At the high energy end, the beam has to be transported to the target which is usually done with magnetic quadrupole lenses since solenoids and electric lenses are not effective at high particle velocities.

0094-243X/86/1390045-17$3.00

In an HIF system, this transport has to be provided over a very large distance, and a periodic FODO channel is required to accomplish this task. The major bottleneck in an HIF accelerator though is at the low energy end. Since the currents produced by ion sources are several orders of magnitude below the levels required for pellet ignition, various schemes of current multiplication have to be employed such as longitudinal pulse compression, multiple-beam arrays in the induction linac, or funneling in the rf linac system for HIF.

The first systematic studies of high-intensity, high-brightness beams occurred in the late sixties and early seventies in connection with the design and operation of the high-energy injector linacs at CERN, Brookhaven, and Fermilab and the LAMPF accelerator at Los Alamos. Particularly noteworthy for this period is the development and use of computer simulation codes[1] to study the particle acceleration and transport through an rf linac. In addition, considerable progress was made in developing analytical tools to describe the effects of space-charge on the beam dynamics, such as the concept of rms emittance and equivalent beams by Lapostolle and Sacherer[2].

The main results of both computer simulations as well as measurements from that period can perhaps best be summarized by the following empirical relation between output and input emittance (normalized) for an rf linac:

$$\varepsilon_{N,out} = (\varepsilon_{N,in}^2 + kI^n)^{1/2} , \qquad (1)$$

Here, I is the current in the bunch, n a number in the range $0.6 < n < 1.0$ and k a constant. This relationship, deduced from computer simulation results by R. Chasman and Promé[3] implies that there is a fundamental limit to the emittance and brightness of a rf linac beam which depends on the beam current. As the input emittance is decreased ($\varepsilon_{N,in} \to 0$), the output emittance approaches a lower limit $\varepsilon_{N,out} = (kI^n)^{1/2}$ which depends on the current. It should be emphasized that there are to date no theoretical models that explain the linac beam physics responsible for the above relation and that predict the exact values for the parameters k and n.

The proposals by Maschke and Martin to use heavy ion accelerators for inertial fusion triggered new interest in the transport of intense beams and launched the present period of research activity in this field that began in the late seventies. In the meantime, of course, other applications of high-brightness beams have broadened the scope of this work considerably. The main focus initially has been the beam transport in a long periodic channel where theoretical studies predicted instabilities due to resonant interaction between beam modes and the periodic structure.

Looking back at the past few years, one can say that the concentration on transverse beam transport of long beams proved to be very beneficial. The absence of longitudinal effects led to the successful diagnosis of various phenomena that tend to be obscured by the complicated three-dimensional beam dynamics in an rf linac. An excellent example is the effect of emittance growth in nonuniform beam distributions via conversion of field energy into transverse kinetic energy first identified in computer simulation studies at GSI Darmstadt[4] and discussed in the paper by Wangler et al.[5]. The emittance growth formula in this case is very similar to Eq. (1), and one can expect that this effect plays also an important role in bunched beams.

The following review of beam physics issues and ongoing experiments is limited for the most part to transverse focusing (transport) of long beams. Although the emphasis of recent work has been on periodic focusing, we will take a broader view here and include other aspects such as the particle source, transport through individual lenses or drift tubes, and charge neutralization effects.

CURRENT AND BRIGHTNESS LIMITS OF SOURCES AND TRANSPORT SYSTEMS

The central issue for the design of a high-intensity, high-brightness accelerator is that a beam with a desired current and emittance be delivered to a target without appreciable loss of particles and without excessive emittance growth during the transport and acceleration process. In this section we will discuss the fundamental limits on current and brightness imposed by the particle source and various focusing elements in a low-energy beam transport system.

The typical source of electrons is a thermionic cathode while ions are usually extracted from the plasma of a gas discharge. When the cathode or the plasma are in thermal equilibrium at a temperature T, the velocity distribution of the particles is Maxwellian, i.e., $f(v) = \exp(- mv^2/kT)$. The source temperature represents an intrinsic lower limit for the normalized emittance, ε_N. If r_s denotes the radius of the cathode in an electron gun or of the emitting plasma surface of an ion source, and m_oc^2 the rest energy of the particles, the intrinsic normalized emittance is given by the relation[6]

$$\varepsilon_N = \beta\gamma\varepsilon = 2\, r_s(kT/m_oc^2)^{1/2} , \qquad (2)$$

where $\beta = v/c$, $\gamma = (1 - \beta^2)^{-1/2}$. In electron guns kT is typically about 0.1 eV; for plasma ion sources kT may be several times larger than that.

The current density of the particle beam extracted from a source is limited intrinsically by the Child-Langmuir law (in MKS units):

$$J = 1.67 \times 10^{-3} \left(\frac{q}{m_oc^2}\right)^{1/2} \frac{V_o^{3/2}}{d^2} . \qquad (3)$$

V_o is the extraction voltage and d the spacing between the emitter surface and the extraction electrode. This implies a fundamental current limit of

$$I = 0.52 \times 10^{-2} \left(\frac{q}{m_o c^2}\right)^{1/2} \left(\frac{r_s}{d}\right)^2 V_o^{3/2} . \quad (4)$$

From Eqs. (2) and (4), one obtains an upper bound for the normalized brightness $B_N = I/\varepsilon_N{}^2$ of a charged particle beam.

In practice, one cannot achieve the high currents predicted by the Child-Langmuir law. Current densities for cathodes used in electron guns, for instance, are limited to about 10 A/cm^2. On the other hand, electrical breakdown, imposes limits to the values of the voltage V_o and gap spacing d, and hence the current, that can be achieved in any source. Keller states this breakdown limit in the empirical relation[7]

$$d = 1.4 \times 10^{-3} \, V_o^{3/2} , \quad (5)$$

where V_o is in kV and d in cm. Furthermore, to avoid nonlinearities, and hence emittance deterioration, in the ion optics of the source, the aspect ratio between the emitter radius r_s and the electrode spacing d must stay within limits which according to Keller are typically in the range[7]

$$0.5 \leqslant \frac{r_s}{d} \leqslant 1.0 . \quad (6)$$

The above relations define the intrinsic limits imposed by the source on the emittance, the current, and the brightness of a particle beam. They can serve as a benchmark for computer simulations or experimental data. In practice, one finds that the normalized emittance of the beam produced by an accelerator may be many times larger than the intrinsic limit given by Eq. (2).

Many high-intensity ion sources are capable of producing higher beam currents than can be handled in a beam transport system or by the low-energy end of the linear accelerator. Thus, it is important to know the theoretical current limits of various focusing systems such as a short lens followed by a drift tube, a long solenoid, a periodic focusing channel, and neutralization of the space charge. The current-handling capability of each focusing system can best be evaluated by using the linear K-V beam model in which the space-charge forces as well as the external forces acting on the particle are proportional to the distance of the particle from the beam axis. The equation for the envelope radius r = a of such a beam has the well-known form

$$a'' + \kappa a - \frac{K}{a} - \frac{\varepsilon^2}{a^3} = 0 , \quad (7)$$

where κ denotes the focusing strength, ε the unnormalized emittance, and K the generalized perveance[8] defined as

$$K = \frac{I}{I_o} \frac{2}{\beta^3 \gamma^3} . \tag{8}$$

The characteristic current I_o is given by $I_o = 4\pi\varepsilon_o m_o c^3/q$ and has the values 1.7×10^4 amperes for electrons and 3.1×10^7 (A/Z) amperes for ions of mass number A and charge $q = Ze$.

In the nonrelativistic approximation, where $\gamma \approx 1$, $\beta \approx (2qV/m_o c^2)^{1/2}$, one obtains the relations

$$K = 1.515 \times 10^4 \frac{I}{V^{3/2}} \text{ (for electrons)} \tag{9}$$

and

$$K = 6.474 \times 10^5 \frac{I}{V^{3/2}} \left(\frac{A}{Z}\right)^{1/2} \text{ (for ions)} , \tag{10}$$

where I is the electrical beam current measured in amperes and V the voltage equivalant of the kinetic energy measured in volts.

In high-brightness beams without charge neutralization, the effect of the emittance on the beam radius is negligible, i.e. one can assume that $\varepsilon^2/a^3 \ll K/a$; the generalized perveance K takes on the role of a scaling parameter in this case. Let us now consider a transport element with linear aperture radius a and length S. The "linear aperture" is the fraction of the aperture of a lens for which the forces are linear to within an acceptable limit (typically a few percent). The maximum current that can be transported through such an element can be found by solving the K-V envelope equation with $\varepsilon = 0$, which implies that the space-charge force is exactly balanced by the focusing force at every position within the beam.

As a first example, we will take a thin focusing lens followed by a drift tube with radius a and length S. If f denotes the focusing length of the lens, then the maximum current is passed throught the tube when $f = S/3.974$, and the corresponding value of the generalized perveance is[9]

$$K_{max} = 2.33\ (a/S)^2 , \tag{11}$$

The equivalent beam current is

$$I_{max} = 1.165\ I_o \beta^3 \gamma^3 (a/S)^2 . \tag{12}$$

The maximum transportable current is thus proportional to the geometry factor $(a/S)^2$ and to $\beta^3\gamma^3$. For a nonrelativistic ion beam, the maximum current is given by

$$I_{max} = 3.6 \times 10^{-6} \left(\frac{Z}{A}\right)^{1/2} V^{3/2} \left(\frac{a}{S}\right)^{2} \ . \tag{13}$$

As a second example we consider the uniform focusing channel consisting of a long solenoid. In this case one obtains from Eq. (7) with $a'' = 0$, and $\varepsilon = 0$ the result

$$K = \kappa a^2 = \left(\frac{qBa}{2m_0 c\beta\gamma}\right)^2 \tag{14}$$

for the generalized perveance. This value is proportional to the square of the product of beam radius and solenoidal magnetic field B. The theory shows that there is an upper limit for K given by[9]

$$K_{max} = 0.384 \ , \tag{15}$$

which implies a maximum current for a nonrelativistic ion beam of

$$I_{max} = 5.93 \times 10^{-7} \left(\frac{Z}{A}\right)^{1/2} V^{3/2} \ . \tag{16}$$

A third example of beam transport that has received considerable attention in connection with the final transport of the beam in heavy ion fusion is a periodic focusing channel. From the smooth approximation theory one obtains the result[8,10]

$$K_{max} = \sigma_0 \alpha/S = \sigma_0^2 (a/S)^2 G \ , \tag{17}$$

where S is the length of one focusing period, α the channel acceptance, σ_0 is the phase advance per period of the single particle oscillation without space charge, and $G = (\bar{a}/a)^2$ is the ripple factor respresenting the square of the ratio of the average beam radius, $\bar{a}$, to the maximum radius, a. For a FODO channel one finds[11] that $G \approx 1 - \frac{1.2}{\pi}\sigma_0$. Recent theoretical and experimental studies have shown that σ_0 should not be greater than 90^0 and that perhaps $\sigma_0 = 60^0$ is the most prudent choice, i.e. $\pi/3 \leqslant \sigma_0 \leqslant \pi/2$. If one takes the less conservative upper limit ($\sigma_0 = \pi/2$) and considers a quadrupole FODO channel, then one obtains $G \approx 0.4$ and hence

$$K_{max} = 0.987 \ (a/S)^2 \ . \tag{18}$$

Note that the aspect ratio a/S cannot be made arbitrarily large. If one assumes $a/S = 0.5$ as an upper limit, one obtains

$$K_{max} = 0.247 \ . \tag{19}$$

This compares with $K_{max} = 0.583$ for the drift-tube case (assuming the same aspect ratio) and with $K_{max} = 0.384$ for the long solenoid. Thus, in situations where there are no constraints on the focusing magnetic fields, the transportable current is lowest in the periodic channel. It should be noted, however, that a choice between the three options exists only for low-energy transport. At high particle energies, solenoids become impractical and a magnetic FODO system remains the only alternative.

For long beam pulses (say $\tau > 100$ μs) propagating in moderate vacuum ($p \sim 10^{-5}$ Torr), charge neutralization via ionizing collisions with the background gas may occur. Such neutralization may be desirable for high-current transport. But it may also have undesirable effects, e.g. if the degree of neutralization, f, varies within the pulse or is not very reproducible over long time scales. To illustrate the neutralization effect on beam transport, let us assume that the beam is fully neutralized, i.e. $f = 1$. The generalized perveance K in Eq. (7) is then zero, and the radius of the beam is entirely determined by the emittance. From Eq. (7) one finds with $a'' = 0$ and $K = 0$, the solution

$$a_o = (\varepsilon/\sqrt{\kappa})^{1/2} . \tag{20}$$

This radius can be considerably smaller than for the unneutralized beam, i.e. $a_o << a$. Thus, if I_u denotes the current of the unneutralized beam that can be transported through the channel and I_n the current of the neutralized beam, then

$$I_n \approx I_u(a/a_o)^2 . \tag{21}$$

This assumes of course, that more current is available from the source and that the emittance of the neutralized beam remains small enough for the particular application. Charge neutralization is a rather complex issue which is not fully understood yet, and the general attitude has been to avoid it when possible. However, the new interest in high-intensity beams may also lead to more research and better understanding of this problem. We will discuss charge neutralization effects in the last section of this paper.

EMITTANCE GROWTH AND NONLINEAR EFFECTS

The intensity limits of various focusing elements discussed in the previous section apply for ideal systems, i.e. matched beam, linear forces, and low emittance ($\varepsilon << a\sqrt{K}$). However, in practice, many effects may cause emittance growth and beam loss. Major sources of emittance growth are instabilities in periodic focusing channels, nonlinear forces due to external fields or the self fields, exchange between field energy and transverse kinetic energy in nonuniform beams, off-centered beams, lens misalignments and effects of image forces, energy exchange between two directions (equipartioning). We will now briefly summarize the present status and major results of ongoing studies devoted to these issues.

The main effort during the past few years has been concerned with beam transport in a long periodic focusing channel which is of crucial importance for heavy ion fusion. Analytical work and computer simulation using the ideal K-V distribution had indicated[11] that there is a window for safe transport defined by $\sigma_o \leq 60^0$ and $\sigma > 24^0$, where σ and σ_o are the phase advance with and without space charge, respectively. More recent work, both computer simulation[14] with non-K-V distributions as well as experiments, demonstrated that these instabilities do not affect laboratory beams as much as had been feared. Only the envelope-type instability poses an upper limit for σ_o leading to a more relaxed condition of $\sigma_o < 90^0$ with no lower limit on σ. As a result, the transportable current could be almost a factor two higher than had been assumed at most. However, other efforts may impose additional constraints as will be discussed below.

All practical focusing lenses have aberrations; i.e. the force on a particle is not strictly proportional to the displacement from the axis. In axisymmetric lenses, for instance, the spherical aberration adds a cubic term to the radial force, i.e. one has a trajectory equation of the form

$$r'' = - a_1 r - a_3 r^3 \; . \tag{22}$$

For solenoids, the two coefficients a_1 and a_3 are positive; both terms are therefore focusing, and the ratio of the two force terms is $F_3/F_1 = (a_3/a_1)r^2$. Numerical simulation studies for K-V and thermal distributions in a periodic solenoid channel with a nonlinear lens force described by Eq. (22) have shown the following results[12]: at $\sigma_o = 60^0$, there is no emittance growth; at $\sigma_o = 90^0$, on the other hand, significant emittance growth occurs which increases rapidly with F_3/F_1. These nonlinear effects thus may prohibit operation near $\sigma = 90^0$ or reduce the useful aperture radius (and hence acceptance) of the periodic transport channel. We have no experimental confirmation yet of these findings. A systematic study at the University of Maryland of the effects of nonlinear forces on the current density profile of an electron beam passing through a single solenoid lens has been published recently[13]. It was shown that the beam becomes hollow and that a halo forms when the focusing strength is increased.

The above examples illustrate what happens when an initially uniform beam with linear space-charge forces is injected into a system with nonlinear focusing forces. A different situation arises when a nonuniform beam with nonlinear space-charge forces is injected into a linear focusing system. In the above-mentioned simulation studies at GSI we found that considerable emittance growth may occur in this case[4]. As the beam passes through the linear focusing system, the initially nonuniform density distribution tends to become uniform very rapidly, and this redistribution of the particles is associated with an emittance increase. The source for this emittance increase was identified as a conversion of field energy into kinetic energy. A spatially nonuniform beam has a larger amount of stored field energy per unit length than a uniform beam. Near the space-charge limit ($\sigma/\sigma_o \to 0$), theory and simulation show that stationary distributions in a linear channel tend to have a uniform density profile such that the space-charge force is linear everywhere and balances the external focusing force. Thus, when a nonuniform beam is injected into such a linear system it becomes uniform very quickly; in this process the excess amount of field energy is converted into transverse kinetic energy and thus emittance growth. Systematic theoretical and numerical studies by Wangler et al.[5] led to a better understanding of this effect and yielded a differential equation describing the emittance growth in terms of the excessive field energy $\Delta U/w_o$ and the generalized perveance K. From this differential equation one can derive a formula for the emittance growth which is given by

$$\varepsilon_{out} = (\varepsilon_{in}^2 + kI)^{1/2} , \qquad (23)$$

where

$$k = (\Delta U/w_o)(\bar{a}^2/I_o\beta^3\gamma^3) , \qquad (24)$$

It should be noted that the effect occurs in a single lens and does not require a periodic force. Furthermore the effect should be most pronounced in ion beams from plasma sources which tend to have non-uniform (almost Gaussian) density profiles. As will be discussed in the next section, experiments at GSI with an Argon beam appear to confirm this theoretical expectation. On the other hand, electron beams from a disk-shaped thermionic cathode, generally have a rather uniform density profile, and one expects no emittance growth from this effect. This is consistent with observations at the University of Maryland where the electron beam emittance remains within 30% to 40% of the intrinsic value due to the cathode temperature [Eq. (2)] even though the tune depression is very low ($\sigma/\sigma_o \approx 0.1$).

Equation (23) is similar to the empirical relation found for bunched beams in a linac [Eq. (1)], and one expects, therefore, that the above effect may help to explain the emittance growth in linear accelerators.

The problem just described raises a fundamental question for the theoretician, namely how to find stationary distributions in a periodic focusing channel. Until recently, the only known distribution was that of Kapshinskij-Vladimiskij (K-V beam). However, Struckmeier[14] has shown that a generalized waterbag distribution also represents a matched beam in a periodic channel. The associated density profile is nonuniform in general, but becomes uniform as $\sigma \to 0$.

A well-known source of particle loss encountered in experiments is beam off-centering and/or misalignment of lenses. This problem is currently being studied both at Berkeley and at the University of Maryland. An off-centered beam performs coherent oscillations with an effective phase advance given by

$$\sigma_{coh} = (\sigma_o^2 - \sigma_{im}^2)^{1/2} , \tag{25}$$

where σ_{im}^2 is due to the image force seen at the center of the beam, and hence proportional to the beam current. In the electrostatic quadrupole channel at Berkeley the image force due to an off-centered beam contains a sextupole component (cos 3θ). It was found in simulation studies[15] that this force interacts resonantly with a sextupole beam mode which results in emittance growth. Similarly, a dodecapole component (cos 6θ) in the external focusing force (when acting without the image term) was found to result in emittance growth. Fortunately, due to opposite signs in the two forces, one can achieve complete cancellation and suppression of emittance growth by proper choice of the of the quadrupole electrode geometry.

CHARGE NEUTRALIZATION EFFECTS

When a charged particle beam is injected into an evacuated drift tube the space charge creates an electric field in the interior region of the tube. The associated electric potential distribution can be calculated by solving Poisson's equation. The simplest case is that of an axisymmetric beam with uniform density and slowly varying beam envelope where one can find the radial electric field and the potential V(r) readily by applying Gauss' law, $\int \vec{E} \cdot \vec{ds} = \int \rho dv$. If a denotes the beam radius and b the radius of the tube one obtains

$$V(r) = V_s(1 + 2 \ln b/a - r^2/a^2) \qquad \text{for } r \leq a \tag{26}$$

and

$$V(r) = V_s \, 2 \ln b/r \qquad \text{for } a \leq r \leq b , \tag{27}$$

where

$$V_s = \frac{I}{4\pi\varepsilon_o \beta c} = \frac{30\ I}{\beta} \tag{28}$$

represents the potential difference between the beam axis (r = 0) and beam edge (r = a) in volts and I is the beam current in amperes. The potential difference between the beam axis (r = a) and the conducting wall (r = b) is given by

$$\Delta V = V_o = V_s(1 + 2 \ln b/a) . \tag{29}$$

From conservation of total energy follows that a particle at radius r will have its kinetic energy reduced by an amount $\Delta T = q\,\Delta V(r)$, where $\Delta V(r)$ is the potential difference between r and the tube wall (r = b).

Partial charge neutralization of the beam by particles of opposite polarity can be taken into account by including a neutralization factor f(r) in the derivation of the above formulas. Poisson's equation or Gauss' law can be used to find V(r) provided that the radial variation of f(r) is known. The solution is particularly simple when the primary beam is uniform and f = const. In this case one recovers the above formulas provided that Eq. (28) for V_s is redefined as

$$V_s = \frac{30\ I}{\beta}(1 - f) \ . \tag{30}$$

The actual situation is usually much more complicated, and one needs either experiments or theoretical models of the neutralization process to determine f as a function of position within the beam pulse. The only detailed theoretical attempt to model neutralization processes in a low-energy linear accelerator that I am aware of was carried out by Evans and Warner in 1970/71 in connection with experiments at CERN[16]. Since this work is not well known today, I would like to present a brief review. The two authors studied neutralization effects due to collisional ionization of the residual gas by the beam and due to electron emission from walls relevant for a proton beam with energy between 0.5 and 3.0 MeV, a pulse length of 20 μs, beam current in the pulse of 500 mA, and a background vacuum pressure of $p < 10^{-4}$Torr.

Let us first discuss ionizing collisions of the beam protons in a hydrogen (H_2) background of density n_g. If σ is the total ionization cross section, v the proton velocity, $\phi = J/q = n_i v$ the proton flux per $cm^2 s$, then the number of electron-ion pairs produced per cm^3 in time τ is

$$n_e = \phi n_g \sigma\tau = n_i v\ n_g\ \sigma\tau \ . \tag{31}$$

Assuming that all positive ions are ejected from the beam, that there is no secondary ionization or recombination and that the electrons created in the collisions have zero kinetic energy, one can calculate the neutralization time τ_N when $n_e = n_i$, or f = 1:

$$\tau_N = \frac{1}{n_g\ \sigma\ v} \tag{32}$$

For 500 keV protons in hydrogen gas at $p = 10^{-4}$ Torr, one has $\sigma \approx 5 \times 10^{-7}$ cm^2 and obtains $\tau_N \approx 5.8$ μs. Thus one would conclude that the beam pulse is fully neutralized except for the first 6 μs. However, this model with the assumptions stated above is too simplistic. In particular, as Evans and Warner discuss in detail, the electrons produced by collisional ionization have an energy distribution. This implies that all electrons created at a radius r within the beam will escape to the wall if their kinetic energy is greater than the potential difference ΔV, i.e., if $T > e\,\Delta V(r)$. As a result, the neutralization time τ_N may be considerably longer than the value given in Eq. (32) depending on the wall depth ΔV and hence the beam current. The amount of partial charge neutralization, f, is thus a function of time within the pulse. The trapped electrons $[T < e\,\Delta V(r)]$ will oscillate radially across the beam and into the region between beam edge and wall if their energy is large enough. Indeed, as f increases and the net potential wall becomes smaller, more and more electrons will escape to the wall or penetrate into the region between the primary beam and the wall.

In order to obtain a crude estimate of the electron distribution versus radius, $n_e(r)$, Evans and Warner used a thermal equilibrium model. They assumed that the ion density remains uniform, i.e., n_i = const. for $r \leq a$ and $n_i = 0$ for $a < r < b$. The beam ions thus provide a stationary focusing field for the electrons which can be treated like an external electric field. The secondary electrons from the collisions on the other hand, are assumed to evolve into a Maxwell-Boltzmann distribution, i.e.,

$$n_e(r) = n_e(o) \exp[(eV(r) - eV(o)/kT_e] \ , \qquad (33)$$

where kT_e represents the temperature of the electron gas (i.e., kT_e is proportional to the mean kinetic energy of the electrons) and V(r) is the net potential due to both beam ions and secondary electrons. Strictly speaking such a thermal equilibrium model is valid only when the thermalization time for the electron distribution is comparable to the pulse length of the ion beam. However, even if the thermalization time is longer than the pulse time, as was the case in the CERN studies, the model is still useful in that it gives the equilibrium state towards which the electron distribution tends to go.

The net potential V(r) is found from Poisson's equation, i.e.,

$$\nabla^2 V = \frac{1}{r}\frac{d}{dr}\left(r\,\frac{dV}{dr}\right) = -\frac{e}{\varepsilon_o}(n_i - n_e) = -\frac{en_i}{\varepsilon_o}(1 - f) \ . \qquad (34)$$

By substituting Eq. (33) for n_e into this equation one can solve numerically for V(r) and then also for $n_e(r)$, or f(r). When kT_e is comparable to the potential V_s given by Eq. (28), the electron distribution resembles a Gaussian profile with $n_e(o)$ considerably less than n_i and a large tail between beam and wall. On the other hand, in the limit $kT_e \rightarrow 0$, the electron distribution evolves towards a uniform distribution with $f = 1$ for $r \leq a$ and $f = 0$ for $a < r < b$.

In summary, for the proton beam considered, Evans and Warner concluded that full neutralization ($f = 1$) was not possible due to the finite electron temperature. They also investigated neutralization by secondary electrons knocked out from the walls by beam protons or other particles including photons. In this case, too, the theoretical model predicted that full neutralization could not be achieved.

Evans and Warner also studied the effects of magnetic quadrupole fields on the electron motion and the neutralization process. In this case the electrons perform $\vec{E} \times \vec{B}$ drift, but trapping and hence partial neutralizational is still possible. On the other hand, positively biased electrodes placed along the beam can collect most secondary electrons and prevent charge neutralization of the ion beam. (We note in this context that charge neutralization does not occur when electric lenses, such as electrostatic quadrupoles are employed to focus the positive ion beam.)

Evans and Warner concluded their studies with the following statement: "It is clear that beam dynamics calculations made neglecting neutralization cannot be generalized rigorously to partially neutralized beams simply by the introduction of a global neutralization factor [(1 - f) in the literature] into the computation of the radial space charge field. In view of the difficulties in understanding and hence controlling neutralization we would suggest that the design of focusing channels for intense low-energy proton beams should not rely on this type of neutralization."

The designer of a low-energy focusing system that transports the beam from the ion source to the accelerator is thus faced with a dilemma. On the one hand, charge neutralization effects may allow him to get the full beam current produced by the ion source to the accelerator entrance while in the absence of neutralization a large fraction of the beam may be lost due to Coulomb repulsion. On the other hand, the accelerator may not be able to handle the full beam current and matching may be a major problem, especially if the fractional neutralization varies along the beam pulse. As a result, the beam loss in the front end of the accelerator may be comparable to, or even exceed, the gain due to the use of neutralization effects in the transport channel. Clearly, further studies extending the earlier work by Evans and Warner will be necessary to obtain a better understanding of charge-neutralization processes. In particular, special attention must be given to the problem of matching a partially neutralized beam into the front end of the RFQ or linac where the electrons are almost instantly swept out of the beam by the electric fields. At this point, no general prescription can be given whether it is better to avoid neutralization altogether (e.g., by using electrostatic lenses or bias electrodes) or to employ it in the hope of getting higher beam current into and through the accelerator.

Lastly, we should note that the above review of neutralization effects was concerned with proton (or positive ion) beams in a regime where the product of pulse length τ and gas pressure p is of the order $p\tau \approx 10^{-4} - 10^{-2}$ torr μs. When $p\tau$ is considerably larger than 10^{-2}, as is the case with d.c. beams, secondary ionization, recombination, emittance growth by scattering, etc. come into play and change the beam physics significantly. Likewise, if the primary beam consists of negative ions, such as H^-, the charge neutralization processes are quite different. In this case, the secondary electrons are expelled and the positive ions are trapped. Also, ionizing collisions with the background gas molecules compete with charge exchange processes where H^- becomes a neutral hydrogen atom (H^0) or a proton (H^+). Work on H^- transport is in progress at Los Alamos[17] and other places.

EXPERIMENTS

Several years ago when analytical theory and computer simulation predicted instabilities in long periodic focusing channels there was no experimental data to check the validity of these results. Thus, experiments were launched at the University of Maryland, the Lawrence Berkeley Laboratory, GSI, and Brookhaven to answer some of the major questions raised. Over the past two years these experiments have produced first results. As mentioned above, these results indicate that beam transport in a periodic channel is not as constricted by instabilities as had first been expected. This was also confirmed by recent simulation studies with more realistic particle distributions. On the other hand, the experiments have raised new questions such as nonlinear focusing effects, off-centering, image forces, and charge neutralization that are currently major topics of research. In addition, multiple-beam arrays are being studied at Berkeley and the FOM Institute at Amsterdam. In this last section, I will present a few highlights of the ongoing experiments at the University of Maryland, Berkeley, and GSI.

At the University of Maryland[18,19] an electron beam (5 keV, 50 to 250 mA) is being transported through a periodic solenoid channel. To avoid charge-neutralization effects, the pulse length is only a few microseconds. The first stage of the channel consisted of 2 matching lenses and 12 periodically spaced lenses. We demonstrated 100% transmission of the injected current over a window of $40^0 < \sigma_o < 110^0$, with a tune depression of $\sigma/\sigma_o \approx 0.1$. The intrinsic theoretical emittance of the electron gun is 9×10^{-5}m-rad (unnormalized), the measured emittance was 10×10^{-5} m-rad before injection into the focusing system and 13×10^{-5} m-rad at the end of the 12-lens section (for $\sigma_o = 60^0$ and $\sigma_o = 90^0$). Thus, after the transport through 14 lenses, the beam emittance was only about 40% higher than the intrinsic value.

Of particular interest has been the effect of nonlinear lens forces (spherical aberrations) on the beam envelope. In the $\sigma_o = 60^0$ case, the measured beam radius is approximately 10% smaller than the theoretical result based on linear beam optics. P. Loschialpo[20] developed a computer program that includes nonlinear effects in the external focusing forces up to third order as well as the nonuniform space charge. With this program he obtained results that are in excellent agreement with the experimental data indicating that the decrease in the beam radius is entirely due to the third-order force terms which provide a net focusing effect.

Last year we added a second stage of 24 additional lenses to the channel bringing the total to 36 focusing periods. Beam transport studies with a different electron gun from the Rutherford-Appleton Laboratory having a smaller cathode radius (0.5 cm vs. 1.25 cm of our large gun) yielded 100% beam transmission through the full channel for $40^0 \leq \sigma_o \leq 100^0$ with rapid beam loss on either side of this window. Following these studies we put our large gun back into the system (after changing the cathode that had deteriorated). We found that the beam was off-centered, largely due to misalignment of the gun, and we obtained 100% transmission only in a narrow range of σ_o values near $\sigma_o = 70^0$. Details of these studies are reported in our paper at the 1985 Particle Accelerator Conference[19]. Since off-centering and image forces are of vital importance, we decided to embark on a systematic study of these effects. We hope that the experimental conditions can be studied with a modified computer simulation code.

The beam transport experiment at Berkeley[21] consists of a Cs^+ source (120-160 keV, 0.7-23 mA), a matching section of 5 electrostatic quadrupoles and a periodic focusing channel with 82 electrostatic quadrupoles (41 periods). Current-attenuating grids and emittance-varying grids are used to cover a wide range in σ_o, σ parameter space. The major result of detailed experimental studies is that beam transport was found to be stable below $\sigma_o = 90^0$ and down to accessible values of the depressed phase advance as low as $\sigma = 8^0$. Unstable regimes of beam transport above $\sigma_o = 90^0$ were found to be in agreement with theoretical expectations. The emittance of the beam from the source is about 5 times larger than the intrinsic value according to Eq. (2)--probably due to nonuniform emission[22]--but remains remarkably constant in the stable regime of transport through the channel.

In recent months the first stage of a multiple-beam experiment with four parallel beams, called MBE-4, was put into operation. It consists of 4 Cs^+ beams (200 keV, 5-10 mA design goal) in electrostatic quadrupole arrays between induction linac acceleration gaps. The final design energy is 1.0 MeV and longitudinal compression will amplify the current by a factor four. This project is designed to study beam physics issues relevant to heavy ion fusion, in particular to develop a data base for the proposed 16-beam high-temperature experiment (at a beam energy of 125 MeV and final current of 96 A). Details of the Berkeley program can be found in the invited paper by D. Keefe at the 1985 Particle Accelerator Conference[23].

At the GSI laboratory, a short periodic channel consisting of 12 magnetic quadrupole lenses in a FODO configuration (6 periods) is being used to study the transport of high-brightness Ar^+ beams (190 keV, several mA)[24]. One major issue at Darmstadt is charge neutralization. The ion pulse length of $\tau \geq 0.5$ ms is long enough that partial neutralization occurs which varies along the beam pulse. The pressure of the background gas is between 10^{-6} and 10^{-7} Torr. Another issue is the fact that the ion beam has a nonuniform profile. This may lead to the rapid emittance growth (via transfer of field energy into kinetic energy) described earlier. When the beam is partially neutralized, the nonlinear space-charge force responsible for the emittance growth is reduced by a factor 1 - f, where f is the degree of neutralization. This is indeed observed in the experiments where it is found that the emittance varies along the beam pulse at the channel entrance. In the early part of the pulse, where neutralization is negligible, the emittance is almost a factor two larger than in the remainder of the pulse. The emittance growth after passage of the beam through the FODO channel is also a function of time within the pulse. It is again largest at the beam front, where growth factors of 2-10 are observed depending on beam current, σ_o, and input emittance. However, the neutralized part of the beam pulse also shows appreciable emittance growth the source of which is not fully understood yet. The degree of neutralization is inferred from measurements of the energy of positive residual gas ions that are ejected from the argon beam. It is found that the ion energy decreases from a peak at the beam front reaching a plateau after about 200 μs which remains constant during the remaining part of the pulse. The plateau corresponds to a partial neutralization of about 60%. Thus, the GSI measurements appear to support the conclusions by Evans and Warner that full neutralization is not occuring in the $p\tau$ regime of 10^{-4} - 10^{-2} torr μs applicable to both the CERN as well as the GSI experiments.

REFERENCES

1. R. Chasman, IEEE Trans. Nucl. Sci. 16, 202 (1969); M. Martini, M. Promé, Proceedings of VIIth Int. Conference On High-Energy Accelerators (Yerevan, USSR, 1969), p. 223.
2. P. M. Lapostolle, IEEE Trans. Nucl. Sci. 18, 1101 (1971); F. J. Sacherer, ibid., p. 1105.
3. R. R. Stephens, Proceedings of Workshop on Space Charge in Linear Accelerators, (Los Alamos, 1977), LA-7265-C, p. 11, and private communications.
4. J. Struckmeier, J. Klabunde, and M. Reiser, Part. Accel. 15, 47 (1984).
5. T. P. Wangler, K. R. Crandall, R. S. Mills, and M. Reiser, Proceedings of 1985 Particle Accelerator Conference, to be published in IEEE Trans. Nucl. Sci. NS-32, (1985).
6. J. D. Lawson, The Physics of Charged Particle Beams, (Clarendon Press, Oxford, 1977), p. 221.
7. R. Keller, Proceedings of the 1984 Linear Accelerator Conference, GSI-84-11 (Darmstadt, W. Germany), p. 19.
8. M. Reiser, Part. Accel. 8, 167 (1978).

9. M. Reiser, IEEE Trans. Nucl. Sci. 24, 1009 (1977).
10. M. Reiser, J. Appl. Phys. 52, 555 (1981).
11. I. Hofmann, L. J. Laslett, L. Smith, and I. Haber, Part. Accel. 13, 145 (1983).
12. H. Dantsker-Rudd, I. Haber, M. Reiser, Proceedings of 1985 Particle Accelerator Conference, to be published in IEEE Trans. Nucl. Sci. NS-32, (1985).
13. P. Loschialpo, W. Namkung, M. Reiser, and J. D. Lawson, J. Appl. Phys. 57, 10 (1985).
14. J. Struckmeier, J. Klabunde, and M. Reiser, Proceedings of the 1984 Linear Accelerator Conference, (Darmstadt, W. Germany), GSI-84-11, p. 359.
15. L. Smith, private communication; C. M. Celata, I. Haber, L. J. Laslett, L. Smith, and M. G. Tiefenbach, Proceedings of 1985 Particle Accelerator Conference, to be published in IEEE Trans. Nucl. Sci. NS-32, (1985).
16. L. R. Evans, and D. J. Warner, Space Charge Neutralization of Intense Charged Particle Beams: Some Theoretical Considerations, CERN/MPS/LIN 71-2, August 1971.
17. J. D. Sherman, P. Allison, and H. V. Smith, Proceedings of 1985 Particle Accelerator Conference, to be published in IEEE Trans. Nucl. Sci. 32 (1985).
18. M. Reiser, E. Chojnacki, P. Loschialpo, W. Namkung, J. D. Lawson, C. Prior, and G. P. Warner, Proceedings of the 1984 Linear Accelerator Conference, (Darmstadt, W. Germany), GSI-84-11, p. 309.
19. J. McAdoo, E. Chojnacki, P. Loschialpo, K. Low, M. Reiser, and J. D. Lawson, Proceedings of 1985 Particle Accelerator Conference, to be published in IEEE Trans. Nucl. Sci. NS-32, (1985).
20. P. Loschialpo, "Effects of Nonlinear Space Charge and Magnetic Forces on Electron Beams Focused by a Solenoid Lens," Ph.D. thesis, University of Maryland, 1984.
21. A. Faltens, D. Keefe, C. Kim, S. Rosenblum, M. G. Tiefenback, and A. Warwick, Proceedings of the 1984 Linear Accelerator Conference, (Darmstadt, W. Germany), GSI-84-11, p. 312; M. G. Tiefenback and D. Keefe, Proceedings of 1985 Particle Accelerator Conference, to be published in IEEE Trans. Nucl. Sci. NS-32, (1985).
22. C. Kim, private communication.
23. D. Keefe, Proceedings of 1985 Particle Accelerator Conference, to be published in IEEE Trans. Nucl. Sci. NS-32, (1985).
24. J. Klabunde, A. Schönlein, P. Spädtke, Proceedings of 1985 Particle Acceleator Conference, IEEE Trans. Nucl. Sci. NS-32, (1985).

HIGH-CURRENT BEAM TRANSPORT WITH MULTIPLE BEAM ARRAYS*

Charles H. Kim
Lawrence Berkeley Laboratory, University of California
Berkeley, California 94720

ABSTRACT

Highlights of recent experimental and theoretical research progress on the high current beam transport of single and multiple beams by the Heavy Ion Fusion Accelerator Research (HIFAR) group at the Lawrence Berkeley Laboratory (LBL) are presented. In the single beam transport experiment (SBTE), stability boundaries and the emittance growth of a space charge dominated beam in a long quadrupole transport channel were measured and compared with theory and computer simulations. Also, a multiple beam ion induction linac (MBE-4) is being constructed at LBL which will permit study of multiple beam transport arrays, and acceleration and bunch length compression of individually focused beamlets. Various design considerations of MBE-4 regarding scaling laws, nonlinear effects, misalignments, and transverse and longitudinal space charge effects are summarized. Some aspects of longitudinal beam dynamics including schemes to generate the accelerating voltage waveforms and to amplify beam current are also discussed.

I. INTRODUCTION

During the past decade there has been considerable interest in high current beam transport and the stability of space charge dominated beams in accelerators, because of their potential applications to various fields such as inertial confinement fusion [1] [2] , radiation testing facilities [3], neutron spallation sources [4], and other industrial applications [5]. In these accelerators, one seeks to have the space charge defocusing force almost equal to the average channel focusing force. Transverse and longitudinal beam containment and the stability of the beam is of major concern.

In most high current accelerators, transverse containment is achieved by alternating gradient (AG) focusing [6]. The beam envelope equations for AG focusing were derived first by Kapchinskij and Vladimirskij [7] using a rather specialized particle distribution -- a uniform distribution on the surface of a four dimensional sphere in transverse phase space (K-V distribution). Later, Lapostolle [8] and Sacherer [9] showed that the rms envelope equations for certain other distributions are identical to the K-V envelope equations if the rms emittances are used. The scaling laws derived from the envelope equations and some of their consequences are discussed in section II.

---

* This work was supported by the Office of Energy Research, Office of Basic Energy Sciences, Department of Energy under Contract No. DE-AC03-76SF00098.

The Single Beam Transport Experiment (SBTE) was constructed in 1983 to test the stability of a space charge dominated beam in a long transport channel. It consists of 87 electrostatic quadrupole lenses, each of which is 10 cm long and 5 cm aperture diameter. The lattice half period is 15 cm. Typical beam parameters of the cesium +1 beam are : kinetic energy 120 keV, variable current (up to 13 mA), and variable normalized emittance (1.2 - 6.0 x 10-7 pi rad . m). [10]*

An ion induction linac called the Multiple Beam Experiment (MBE-4) is being constructed at LBL [11] to test multiple beam transport arrays [12], and demonstrate acceleration and bunch length compression of space charge dominated ion beams [13]. Construction is scheduled to be completed late in 1986. It has 30 lattice periods each of which is 45.72 cm long and has an aperture diameter of 5.4 cm. It has 24 accelerating gaps; After each set of 4 accelerating gaps, the 5-th gap is left for diagnostic access. The 4-beam ion injector is presently operational [14]. Cesium ions, 200 keV, variable in current up to 15 mA per beamlet, and normalized emittance 1.2 x 10^{-7} pi rad.m will be injected into the beam conditioning unit composed of 8 matching quadrupoles. A final energy of 800 keV, a current amplification of factor 3, and a spatial bunch compression of factor 1.5 are anticipated.

The transportable current limit in a channel increases as the beam is accelerated. The length of the bunch needs to be compressed in a carefully programmed fashion to exploit the increasing current-carrying capability of the channel. This topic is discussed in sections III and IV.

II. TRANSVERSE DYNAMICS OF A SINGLE BEAM

II.1 Scaling

Scaling laws for the current per channel (I), the maximum beam radius (a), and the length of the half lattice period (L) were obtained by Lambertson, Laslett, and Smith by transforming the K-V envelope equations into dimensionless forms [15]. The scaling laws for a matched beam is listed below:

$$I = 7.86 \times 10^6 \ (A/q)\ (\Theta^2 Q/u^2)\ (a\ /L)^2\ (\beta\gamma)^3 \tag{1}$$

$$a^2 = \frac{u^2}{\Theta}\ \frac{\varepsilon_n L}{\beta\gamma} \tag{2}$$

$$L = \Theta\sqrt{\frac{(B\rho)}{B'}} = \Theta\sqrt{\frac{a_Q^2\ V}{V_Q}} \tag{3}$$

where A is the atomic mass number, q charge state, β, γ, the relativistic parameters, ε_n normalized beam emittance, $(B\rho)$ beam momentum, B' magnetic quadrupole field gradient (B_Q/a_Q), a_Q aperture of the channel,

*Throughout this paper emittance is used in the sense of four times the rms normalized emittance.

V_Q voltage on electrostatic quadrupole, and V is the kinetic energy of the beam in eV divided by the charge state q. The parameters Θ, u and Q appear in the dimensionless envelope equations and the values have been tabulated by Laslett [16] as functions of σ and σ_0, where σ and σ_0 are the phase advance per lattice period of the betatron oscillations with and without the space charge defocusing forces. Some useful approximations to these values for electrode occupancy factor of 0.5 are shown below:

$$\Theta^2 Q / u^2 = 6.95 \times 10^{-4} \sigma_0^{3/2} (1 - \sigma^2 / \sigma_0^2) \qquad (4)$$

and

$$\Theta = 0.248 \sigma_0^{0.45} \qquad (5)$$

Eq. (4) is accurate within 2% for $40° < \sigma_0 < 80°$, and $\sigma/\sigma_0 < 0.2$. Eq. (5) is accurate within 1% for $\sigma_0 < 90°$. Similar approximate formulas were derived analytically by Reiser [17] and Lee, Fessenden, and Laslett [18].

Eqs. (1), (2), and (3) can be written in many convenient forms by choosing other relevant parameters as independent variables; for example, if the pole tip field, the beam emittance, and the kinetic energy are chosen as independent variables, Eq. (1) becomes functionally similar to the Maschke formula [19] but with a different numerical coefficient.

II.2 Stability

Until the late 1970's, stable beam transport was believed to be possible only if $\sigma/\sigma_0 > 0.4$ and $\sigma_0 < 60°$, based on the theoretical stability analysis for K-V beams [20]. This restriction had been one of the most important guide lines for designing high current accelerators during this period. During the last several years, experimental investigations and computer simulations using more realistic distributions have shown that space charge alone does not cause any emittance growth if $\sigma_0 < 90°$; only when strong space charge is coupled with misalignments and image charge effects, or with misalignments and lens nonlinearities, does the beam emittance grow [12] [28]. Rapid emittance growth followed by a saturation was predicted theoretically [21] [22], and, in deed, was observed experimentally if the particle distribution is not spatially uniform [37]. Some of the consequences of these new developments on the design of high current accelerators are discussed in this and the following section.

Effects of nonlinearities and misalignments are negligible in SBTE [23]. The initial distribution is approximately semi-Gaussian (flat in configuration space and Gaussian in velocity space) -- a typical feature of thermionic ion sources. In the space charge dominated regime ($\sigma/\sigma_0 \ll 1$), the semi-Gaussian nature is approximately preserved in the transport channel; in the emittance dominated regime ($\sigma/\sigma_0 > 0.5$), the spatial distribution is transformed in the channel and becomes more peaked [24].

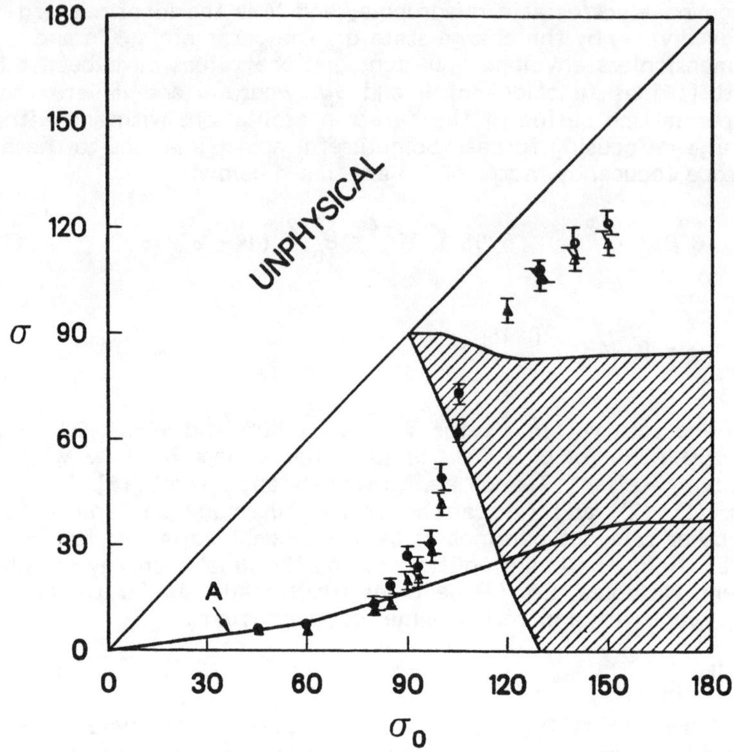

XCG 866-233

Fig. (1) The measured stability boundaries in σ - σ_0 space. The region above the line marked "A" is accessible in SBTE. The envelope equation gives unstable solution in the shaded area [20]. (from Ref. 25)

The stability boundaries in σ - σ_0 space are discussed by M. Tiefenback and D. Keefe [25] and the results are reproduced in Fig. (1). A stable beam was inferred if the current and the emittance both remained unchanged in passing through the entire transport channel of 82 quadrupoles. These measurements largely agree with particle simulations using a semi-Gaussian distribution [26]. The simulations also showed stable beam transport for even smaller values of σ such as $\sigma = 1.5°$ and $\sigma = 60°$.

An emittance growth was observed experimentally in SBTE in the normally stable region (e.g., $\sigma = 8°$, $\sigma = 60°$) when a deliberate misalignment was introduced. The result agrees with computer simulations with image charge forces [12] and will be discussed later in section III.1.

II.3 Adjustments of Focusing Channel as the Beam is Accelerated

Maximum transportable current increases with kinetic energy, but the exact functional dependence on V is different in different limiting regimes discussed below. We assume here that the channel radius does not increase with kinetic energy for the simplicity of discussion.

(1) Low Energy Limit

For low energy beams, the quadrupole voltages are low and the transportable current limit is determined by the beam aspect ratio, a/L [15]. The value of a/L should be kept smaller than about 0.1 in order to avoid nonlinear effects. In this regime, the beam current scales approximately as $\sigma_0^{3/2}$ as the quadrupole voltage is increased.

Thus the maximum transportable current per channel is limited by beam dynamics and not by any technological limits in this regime. For example, for a 1 MeV proton beam the maximum current is about 4 Amperes per channel. For higher total current, transverse stacking of many channels is required [15].

As the beam is accelerated in this regime, it is convenient to keep a and L constant and increase V_Q proportionately with V so that σ is held constant. Since $u \propto (1/\sigma)^{1/2}$, σ decreases as $\beta\gamma^{-1}$. If $\sigma/\sigma_0 \ll 1$, the transportable current increases as $V^{3/2}$. Appropriate longitudinal compression is necessary with the line density increasing as V to take advantage of the scaling.

(2) Breakdown limit

As the beam is accelerated further, the quadrupole voltage can not be increased indefinitely because of the technological limits. In this regime, the simplest configuration of the transport channel for an accelerating beam is to keep a and V_Q constant and increase L as $V^{1/2}$ to keep σ_0 constant. The value of σ does not change and the transportable current scales exactly as $V^{1/2}$, leaving no room for longitudinal bunch length compression.

Varying the length of the quadrupoles continuously along the linac is not very convenient, but if L as well as a and V_Q are kept constant, the maximum transportable current does not increase with kinetic energy.

(3) Magnetic Focusing and the Pole-Tip-Field Limit

For still higher kinetic energies, magnetic quadrupoles are more effective. Depending on the available maximum field values, the transition occurs around 0.1 - 0.5 MeV/amu. In ion induction linacs for HIF application, magnetic focusing consists most of the linac.

Again, if we let the value of σ vary, keep the maximum beam radius fixed, and increase L as $V^{(1/4)}$ to keep σ_0 constant, then the transportable current scales as V and the line density as $V^{(1/2)}$. The value of σ decreases approximately as $V^{-1/4}$.

(4) Tune-Depression Limit

In a transport channel where misalignments and lens nonlinearities are not negligible, there may be a lower bound for the allowable values of σ. If the aperture size and the magnetic field is held constant, L should be increased as $V^{1/4}$ to keep σ_0 constant. If the current is increased as $V^{3/4}$, a decreases as $V^{-1/4}$ and σ remains constant.

A faster current amplification can be achieved if the aperture size is progressively decreased with the kinetic energy and the pole tip field strength is kept constant [27].

II.4 Δβ/β limit

If current amplification is required, the velocity of the tail is necessarily higher than that of the head of the bunch at a given location. If the quadrupole voltage does not vary in time at the given location, the energy and the zero-current-tune (σ_0) dependences of the beam current in Eq. 1 approximately cancel each other and thus the beam radius scales as $\sqrt{I}$. Therefore, an acceleration and bunch compression scheme which gives a constant current at a given location is convenient. This will be discussed in section IV.

Under these circumstances, the maximum beam radius remains constant for a constant current at a given location, but σ, σ_0 , and the ellipticity of the beam shape will vary from the head to the tail of the bunch. These variations may cause mismatch oscillations and coherent betatron oscillations (if there are misalignments) which may cause particle loss if Δβ/β exceeds a certain maximum value.

III. TRANSVERSE DYNAMICS OF MULTIPLE BEAMS

In designing a multiple beam array, it is important to minimize the transverse dimensions in order to reduce the cost of the accelerator. The transverse dimension is determined by many considerations such as alignment tolerances, nonlinearities of the focusing field, image charge effects, beam-beam interaction, and others of a practical nature. In the present summary, we will consider only those effects which were important in designing the MBE-4 multiple beam array (Fig. 2).

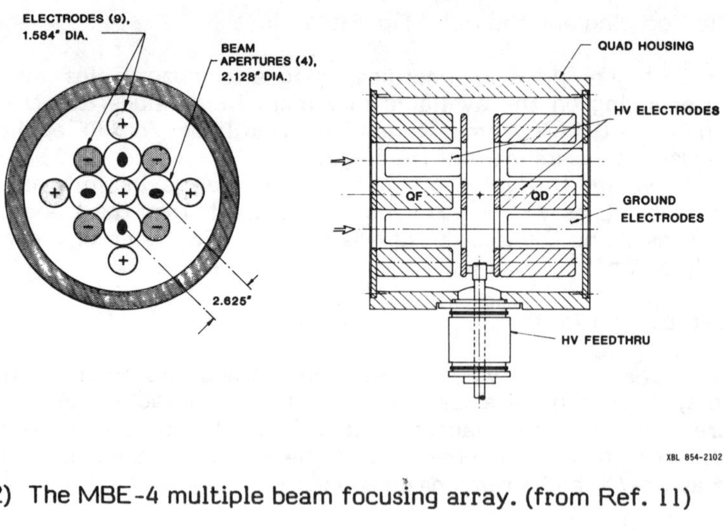

Fig. (2) The MBE-4 multiple beam focusing array. (from Ref. 11)

III.1 Image Charge Effects

Effects on transverse beam dynamics of induced charges on the focusing electrodes was studied using a particle simulation code [12]. For a perfectly aligned beam, image charges did not cause any undesirable effects to the beam, unless the beam radius exceeded 85% of the aperture. For such large beams, particle loss occurred [26].

However, when the beam was misaligned and the tune was sufficiently depressed ($\sigma/\sigma_0 < 0.1$), an unacceptable level of emittance growth was observed (Fig. 3). The emittance growth was due to the excitation of coherent sextupole oscillations driven by the image forces. [12]

III.2 Geometric Nonlinearities

Nonlinear field strengths of the MBE-4 multiple beam focusing arrays were calculated by relaxation codes for use in the simulation calculation. Three classes of nonlinearities were studied : (1) imperfections in the field of a single quadrupole, (2) nonlinearities due to the asymmetry of the geometry with respect to the beamlet of interest, and (3) end effects of the quadrupoles which are supported in an interdigital fashion. (see Fig. 2)

A dodecapole field is the most dominant nonlinearity of type (1). The dodecapole field vanishes only if the radius of the cylindrical electrode is 1.1457 times the aperture radius [29]. In MBE-4, the ratio was chosen to be 0.744 in order to increase the aperture size without increasing the overall size of the array. Another reason for this special ratio is explained below.

In the particle simulation study, if the beam was perfectly aligned, the dodecapole field did not cause any undesirable effect to the beam. However, if the beam was misaligned, and the tune was strongly depressed ($\sigma/\sigma_0 < 0.1$), and the image charge forces were neglected, an emittance growth similar to the one caused by the image forces was observed. One of the remarkable results of the simulation was that for the chosen electrode size the image forces suppressed emittance growth due to the dodecapole, or vice versa [12]. (See Fig. 3)

Effects on beam dynamics of the asymmetry nonlinearities (mainly a dipole field) and the end effects (mainly an octupole field) were found to be negligible in MBE-4 [28].

III.4 Alignment Tolerance

The effect of random misalignments of quadrupole electrodes was treated statistically by L. Smith [30]. As a result of the misalignments, a coherent betatron oscillation is excited. The probability of the amplitude of the oscillation being less than a value δa is given by:

$$P(\delta a) = 1 - \exp\left(-\delta a^2 / A_{rms}^2\right) , \qquad (6)$$

where A_{rms} is a function of misalignments and the number of lattice periods. Thus the probability is 63% for δa being less than A_{rms} and 95% for δa being less than $\sqrt{3}\,A_{rms}$.

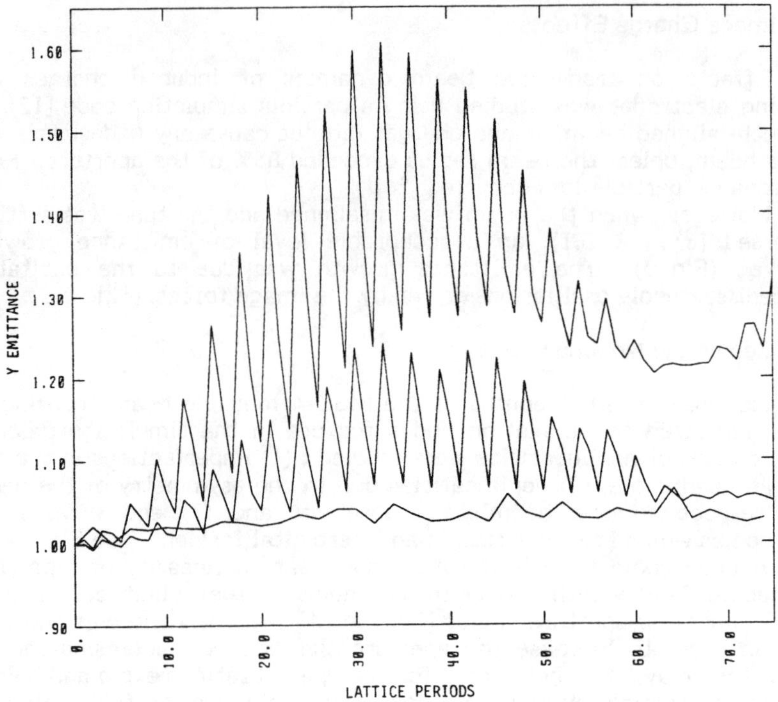

Fig. (3) Calculated emittance growth in MBE-4 for various conditions. The middle curve shows the growth of the beam emittance due to the image charge forces only. Emittance growth due to the dodecapole component only (not shown here) behaves similarly as the middle curve. Addition of a dodecapole field suppressed the emittance growth due to the image charge forces (lower curve). When the polarity of the dodecapole field was reversed (corresponds to larger size electrodes) a stronger emitance growth was observed (top curve). (from Ref. 12)

In MBE-4, a unit consisting of a doublet for each of the four beamlets (see Fig. 2) will be assembled on a bench and brought to the beamline subsequently. In this case, the following three types of misalignment are most serious: transverse misalignments of each lens (Δ); transverse misalignments of doublets (δ); and angular tilt of doublets (α). The functional dependence of Arms on these misalignments is:

$$A_{rms} = \left\{ (11\ \delta_{rms})^2 + (15\ L\ \alpha_{rms})^2 + (22\ \Delta_{rms})^2 \right\} \sqrt{N/30} \qquad (7)$$

where N is the number of lattice periods [30]. The coefficients are relatively insensitive functions of lattice parameters. As an example, if the alignment tolerances are such that $\delta_{rms} = 2\Delta_{rms} = L\ \alpha_{rms} = 0.14$ mm, then $A_{rms} = 3$ mm and the probability of the amplitude being smaller than 10 mm is 99.99%.

For a long wavelength sinusoidal misalignment, a coherent betatron oscillation will be excite resonantly if the wavelength of the misalignment is comparable to that of the coherent betatron oscillation. The coherent oscillation will grow linearly with the distance. The calculated growth rate is 1 mm per wavelength if the misalignment is 0.25mm [31].

III.5 Optimization of the Transverse Dimension of a Multiple Beam Array

In order to minimize the transverse dimensions of a multiple beam array, it is sometimes convenient to maximize the effective current density, I/a_Q^2, where I is the current per channel and a_Q is the aperture radius. This will insure that the overall radius of the array is minimized for a given total current.

We now assume that misalignments and lens nonlinearities are not negligible and that the maximum beam radius and the aperture radius are related as;

$$a + \delta a = \alpha\, a_Q \;; \qquad a_Q > (\delta a/\alpha), \tag{8}$$

where δa provides the required beam clearance due to the misalignments of the channel and α provides the required beam clearance due to the image charge forces, lens nonlinearities, mismatch oscillations and coherent betatron oscillations. ($\alpha \simeq 0.8$)

For each of the transverse focusing regimes discussed in section II.3, one can now find an expression for the effective current density as a function of a_Q only. As an example, consider an electrostatic quadrupole channel operated at the breakdown limit. The breakdown voltage is believed to be related to the aperture size as:

$$V_Q = \text{const}\; a_Q^{\phi} \tag{9}$$

where ϕ is a constant approximately between 0.5 and 0.7 [32]. Eq.(9) can be used to eliminate V_Q in eq.(3), which in turn can be used to eliminate L in eq.(1). By using eq.(8) to eliminate a in eq.(1), one has an expression for I/a_Q

$$I/a_Q^2 = \text{const}\; \frac{Q}{u^2}\, V^{1/2}\, \frac{\left(\alpha\, a_Q - \delta a\right)^2}{a_Q^{(4 - \phi)}} \tag{10}$$

which becomes maximum when;

$$a_Q = \frac{\delta a\;(\,4 - \phi\,)}{\alpha\;(\,2 - \phi\,)}\;, \tag{11}$$

independent of the values of σ, σ_0 and V.

Similar optimizations give slightly different optimum values for different focusing regimes discussed in section II.3. For the aspect ratio limited regime the optimum aperture is $2\delta a/\alpha$; for the magnetic pole tip field limited regime the optimum aperture is $3\delta a/\alpha$.

The value of δa depends not only on the alignment tolerances but also on the frequency of beam steering and on the choice of the acceptable probability of losing the particles. (Eq. 6)

IV. LONGITUDINAL DYNAMICS

As discussed in section II.3, the current carrying capability of the transport channel increases with beam kinetic energy, making room for current amplification. Current amplification is a necessity during acceleration if the capability of the induction linac to accelerate high current is to be exploited. In order to realize current amplification, the tail has to be accelerated more than the head of the bunch in a carefully programmed fashion. It was also pointed out in section II.4 that a constant current at a fixed location is desirable because the beam radius scales as $\sqrt{I}$ in a given quadrupole. Current fluctuations have to be minimized in order to prevent consequent particle losses and longitudinal emittance growth. A procedure to generate accelerating voltage wave forms which satisfy these conditions was devised by Kim and Smith. [33]

IV.1 Current Self-replicating Scheme

High current ion induction linacs have narrow accelerating gaps which are separated by bulky transverse focusing lenses. Both the particle velocity and the line charge density undergo discontinuous jumps as the bunch passes across accelerating gaps. However, the beam current is continuous across each gap.

A current self-replicating scheme is realized if all the particles emerging from an accelerating gap are headed toward a common focal spot in z-t space. The common focal spot is different for different gaps as illustrated in Fig. 4. In this scheme, the shape of the beam current waveform as a function of time at a fixed location is preserved at each position along the linac, as the beam is accelerated and compressed in time. If the current is constant in time at the injection point, it continues to be so at each point along the linac, a desirable feature for transverse focusing as discussed in section II.4. The longitudinal focusing feature is also convenient for certain applications where the bunch is required to be focused to a final focal spot. The space and time dependence of the current, line density, and the accelerating voltage waveforms were calculated analytically for this scheme. [33]

IV.2 Physics Requirements and Technological Constraints

In generating the accelerating voltage waveforms, there are the following physics requirements and constraints of a practical nature.

The requirements dictated by the beam dynamics are: (1) If the beam is not to increase in length on entering the accelerator, a situation which is difficult to overcome by subsequent manipulation of voltage wave forms, the head must not be accelerated before the tail has entered. (2) As explained in section II.4, the value of $\Delta\beta/\beta$ should not exceed a certain upper limit. This relative limit is encountered early in the accelerator, where β is small. (3) Once free of the $\Delta\beta/\beta$ limit, with β continuing to increase, one can use up the available margin in $\Delta\beta/\beta$ to initiate spatial compression of the bunch, by making the acceleration rate of the tail greater than that of the head at a fixed time. The resulting current amplification should follow the required energy dependence described in section III.3.

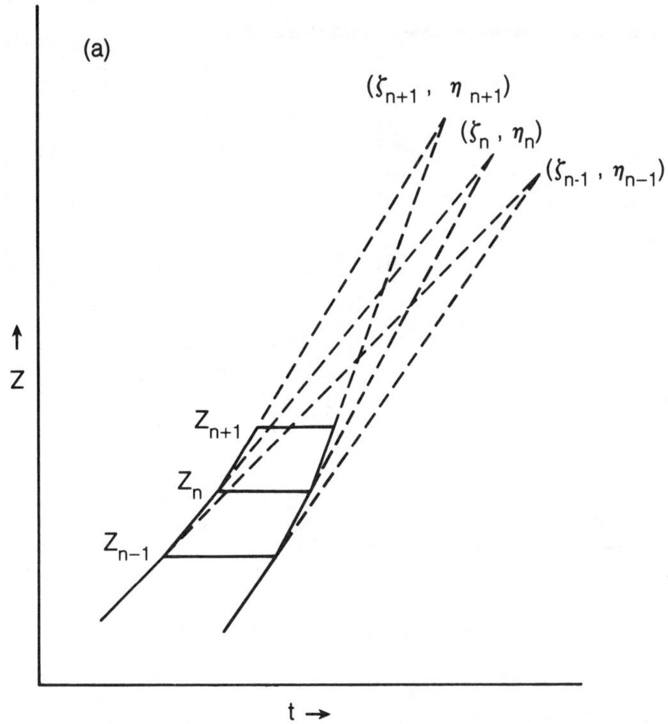

Fig. (4) An illustration of the current self-replicating scheme. Particles emerging from an accelerating gap are headed to a common focal spot. (From Ref. 33)

There are also constraints of a practical nature: (1) The flux swing of the magnetic material of an accelerating module should not exceed a certain upper limit determined by cost. (2) The acceleration rate should not exceed the high voltage breakdown limit. [33] [34]

These requirements and constraints provide the necessary and sufficient conditions for determining the head and tail accelerating voltages at each gap ; thus the common focal spot for each gap.

IV.3 Synthesis of the Accelerating Voltage Waveforms

MBE-4 can be operated in a wide range of beam parameters. An illustrative set of theoretically desired accelerating voltage waveforms for MBE-4 is shown in Fig. 5. These waveforms were approximately synthesized in a numerical study [13] by adding elementary waveforms which were similar to those obtained in laboratory tests [35] [36]. They are curves of positive and negative initial curvatures and an almost flat-top curve with finite rise and fall times.

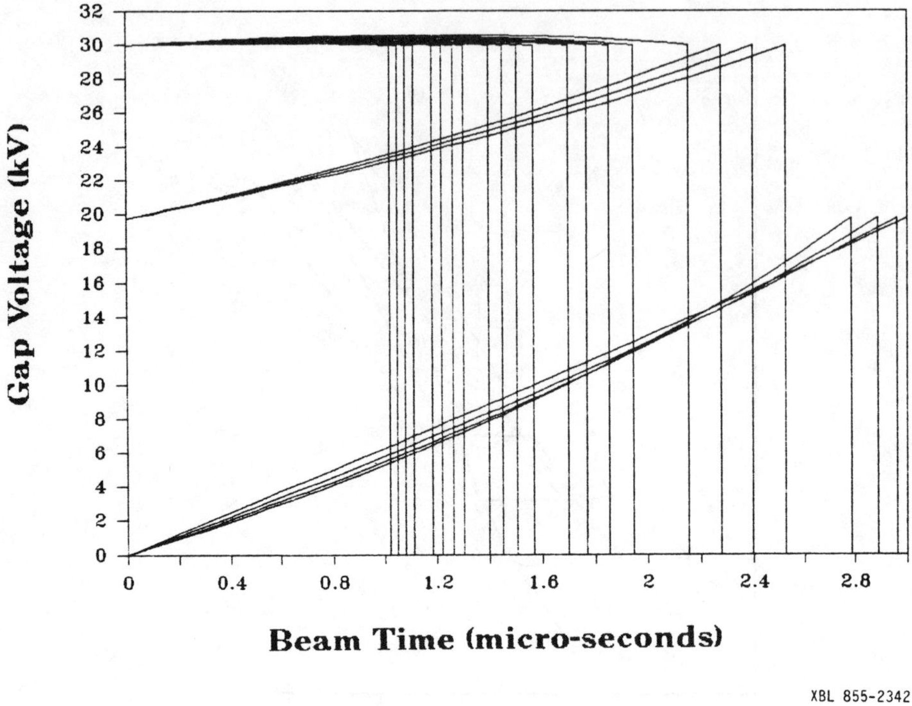

Fig. (5) An example of ideal accelerating voltage waveforms of MBE-4. (from Ref. 13)

A computer code was used to study the effects of the imperfect waveforms on the longitudinal beam dynamics in the presence of strong space-charge forces [13]. Current and kinetic energy errors which are reproducible from pulse-to-pulse can be corrected in principle to an arbitrary accuracy ; however, fluctuations due to timing and voltage jitter cannot be corrected easily. The effects of jitter depend not only on the magnitude of the jitter but also on the way the waveforms are synthesized.

V. SUMMARY

The single beam transport experiment has shown that stable beam transport is possible in the highly space charge dominated regime where the space charge defocusing force is almost equal to the average channel focusing force. However, particle simulations showed that misalignments and the relative size of the beam to the aperture radius played important roles in the stability of space charge dominated beams.

A multiple beam ion induction linac (MBE-4) is under construction at LBL, which will permit study of three dimensional beam dynamics of space charge dominated beams as they are accelerated and compressed longitudinally. It will also permit study of effects of misalignments, image charge forces, and geometric nonlinearities on the stability and emittance growth of the beams.

ACKNOWLEDGEMENTS

The author is indebted to Dr. L. Jackson Laslett for enlightening discussions regarding scaling laws and also for providing tables of calculated values of Θ, Q, and u. The author is also grateful to Drs. C. Celata, D. Keefe, L. Smith, and M. Tiefenback for providing various materials used in this paper, and reading and commenting on the manuscript.

REFERENCES

(1) D. Keefe, Am. Rev. Nucl. Part. Sci., 32, 391(1982), and references therein.

(2) R. Bock, I. Hofmann, and R. Arnold, Nucl. Sci. Applications, 2, (1984) to be published, and references therein.

(3) For example, see R.A. Jameson, IEEE Trans. Nucl. Sci., NS-26, 2986 (1979).

(4) For example, see G.H. Rees, IEEE Trans. Nucl. Sci., NS-30, 3044 (1983).

(5) For example, see M. R. Cleland, Proc. of the Linear Accel. Conf., Seeheim, Fed.Rep. of Germany, 496 (1984).

(6) E.D. Courant and H.S. Snyder, Ann. of Phys., 3, 1 (1958).

(7) I. M. Kapchinskij and V. V. Vladimirskij, Proc. Internat. Conf. on High Energy Accelerators, CERN, 274 (1959).

(8) P.M. Lapostolle, IEEE Trans. Nucl. Sci., 18, 1101 (1971).

(9) F.J. Sacherer, IEEE Trans. Nucl. Sci., 18, 1105 (1971).

(10) C.H. Kim, Lawence Berkeley Laboratory Internal Report, HI-FAN-160, (1981) unpublished.

(11) R. Avery, C.S. Chavis, T.J. Fessenden, D.E. Gough, T.F. Henderson, D. Keefe, J.R. Meneghetti, C.D. Pike, D.L. Vanecek, and A.I. Warwick, IEEE Trans. Nucl. Sci., NS-32, (1985) to be published.

(12) C. Celata, I. Haber, L.J. Laslett, L. Smith, and M.G. Tiefenback, IEEE Trans. Nucl. Sci., NS-32, (1985) to be published.

(13) C.H. Kim, V.O. Brady, T.J. Fessenden, D.L. Judd, and L.J. Laslett, IEEE Trans Nucl. Sci., NS-32, (1985) to be published.

(14) A.I. Warwick, D. Vanecek, and O. Fredricksson, IEEE Trans. Nucl. Sci., NS-32, (1985) to be published.

(15) G. Lambertson, L. J. Laslett, and L. Smith, IEEE Trans. Nucl. Sci., NS-24, 993 (1977)

(16) L. J. Laslett, Lawrence Berkeley Laboratory Internal Report, HI-FAN-287, June (1985)

(17) M. Reiser, J. Appl. Phys., 52 (2), 555 (1981)

(18) E. P. Lee, T.J. Fessenden, and L.J. Laslett, IEEE Trans. Nucl. Sci., NS-32, (1985) to be published

(19) A. Maschke, Quoted by E.D.Courant, Final report of ERDA Summer Study of Heavy Ions for Inertial Fusion, 72 (1976)

(20) I. Hofmann, L.J.Laslett, L. Smith, and I. Haber, Particle Accelerators, 13, 145 (1983)

(21) J. Struckmeier, J. Klabunde, and M. Reiser, Particle Accelerators, 15, 47, (1984)

(22) T. Wangler, IEEE Trans. Nucl. Sci., NS-32, (1985) to be published. Also in these proceedings.

(23) L.J. Laslett and V.O. Brady, Lawrence Berkeley Laboratory Internal Report, HI-FAN-137, (1981) unpublished.

(24) C.H. Kim, A. Faltens, D. Keefe, S.S. Rosenblum, M.G. Tiefenback, and W.I. Warwick, Proc. of the 1984 Linear Accelerator Conference, Seeheim, 312 (1984).

(25) M. Tiefenback and D. Keefe, IEEE Trans. Nucl. Sci., NS-32, (1985) to be published.

(26) I. Haber, Proc. of the 1984 INS International Symposium on Heavy Ion Accelerators and their Applications to Inertial Fusion, Tokyo, 451 (1984).

(27) L.J. Laslett, Lawrence Berkeley Laboartory Internal Report, HI-FAN-229, (1983) unpublished.

(28) C. Celata, V. O. Brady, L. J. Laslett, and L. Smith, IEEE Trans. Nucl. Sci., NS-32, (1985) to be published.

(29) I.E. Dayton, C. Shoemaker, R.F. Mozley, Rev. Sci. Instr., 25, 488 (1954).

(30) L. Smith, Lawrence Berkeley Laboratory Internal Report, HI-FAN-251, (1984).

(31) D.L. Judd, Private Communications (1984).

(32) R. Keller, Proc. of the 1984 Linear Accelerator Conference, Seeheim, 19 (1984), also in these proceedings.

(33) C.H. Kim and L. Smith, Lawrence Berkeley Laboratory Internal Report, HI-FAN-272, (1985) submitted to Particle Accelerators.

(34) E.P. Lee also considered these characteristic regions independently. Private communications (1984).

(35) D.E. Gough, Private communications (1985).

(36) A. Faltens, M. Firth, S.S. Rosenblum, and D. Keefe, IEEE Trans. Nucl. Sci., NS-30, 3669 (1983).

(37) J. Klabunde and P. Spaedtke, IEEE Trans. Nucl. Sci., NS-32, (1985) to be published.

SCHEME TO FUNNEL ION BEAMS WITH A RADIO-FREQUENCY QUADRUPOLE*

R. H. Stokes and G. N. Minerbo,† MS H817
Los Alamos National Laboratory, Los Alamos, NM 87545

ABSTRACT

We describe a proposed method to funnel ion beams using a new form of the radio-frequency quadrupole (RFQ) structure. This RFQ accepts two bunched ion beams and combines them into a single final beam with interlaced microstructure pulses. It also provides uninterrupted periodic transverse focusing to facilitate the funneling of beams with high current and low emittance.

INTRODUCTION

Several new linear-accelerator applications require output ion beams with very high current and small transverse emittance. The transverse beam brightness (beam current divided by the product of the x- and y-emittances) is often used as the figure of merit, and the design objective is to obtain a very high brightness output beam. When linear-accelerator channels operate near their space-charge limit, one method to increase brightness is to combine the beams from two (or more) space-charge-limited channels. The resultant beam would have interlaced microstructure bunches and would be suitable for further acceleration in a linac operating with twice the frequency. This operation has been called funneling and was first proposed as a necessary part of the rf linac approach to heavy-ion fusion.[1] If the funneling of two beams is accomplished with no transverse emittance growth or beam loss, the resulting beam will have twice the current and twice the brightness of the beams that are combined. Increasing beam brightness through funneling requires no improvement in the ion source or accelerator techniques used before the funneling process.

In some applications, to meet brightness requirements, one could use arrays having multiple beams traversing the whole accelerator system with the final beams all focused to a common target spot. If the required brightness were not too high, funneling to combine multiple beams as early as possible would be especially advantageous in large accelerators where most of the acceleration could then be provided in a minimum-cost single-beam accelerator. In general this single-beam accelerator would be less complex and would require less rf power than designs that carry all the initial beams through the entire accelerator system. A funneling tree to combine four beams is shown in Fig. 1. In practice, funneling to increase beam brightness is a difficult

*Work supported by the US Department of Energy.
†Now with Schlumberger, Houston Engineering Center.

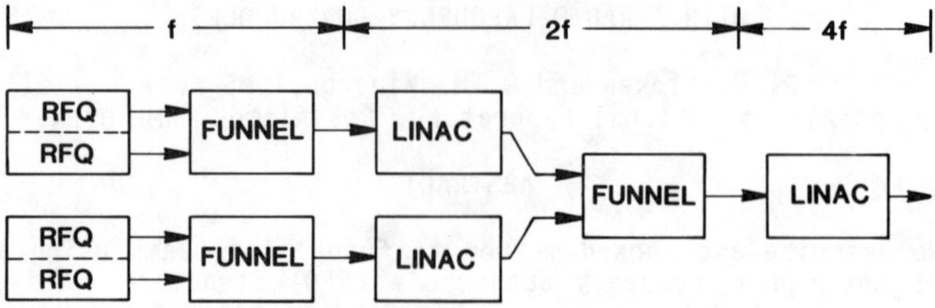

Fig. 1. Funneling tree.

design problem. Initial work using discrete optical elements has been done by Bongardt[2] and by Guy and Wangler.[3]

In 1981, we suggested[4] that ion beams could be funneled with an RFQ of special design. This paper discusses use of this RFQ for the funnel stages (shown in Fig. 1) immediately following the RFQ accelerators. In a funneling tree, the first type of funnel following the RFQ accelerator is present in greater numbers than all other types combined, so it is especially important that a capable and efficient design be developed. Instead of discrete optical elements, we propose to funnel entirely within the special RFQ structure so that the beams are always contained by uninterrupted periodic focusing. Beams with high space charge experience irreversible emittance growth when they emerge from a periodic focusing system or when their focusing periodicity is abruptly changed. To alleviate this problem, in the proposed funneling scheme it should be possible to provide uninterrupted focusing with the same periodicity as that of the accelerators preceding the funnel. Also, instead of conventional deflection systems, we propose to use the properties of the special RFQ structure to deflect two parallel beams toward each other and to merge them into a single final beam. This funneled beam would then be suitable for further acceleration in a linac with twice the frequency of the RFQ funnel. The initial RFQ accelerator, the RFQ funnel, and the following linac would all have every longitudinal bucket filled with ion bunches, and the capability of the accelerator system would be fully utilized.

BEAM DEFLECTION IN PERIODIC FOCUSING CHANNELS

Our first objective was to devise a method to produce transverse deflections of a beam traveling in a periodic focusing system such as an RFQ. We found that introducing periodic transverse displacements of the focusing and defocusing lenses was very effective in producing a deflecting force. To develop this idea, we used the thin-lens array shown in Fig. 2. The input beam

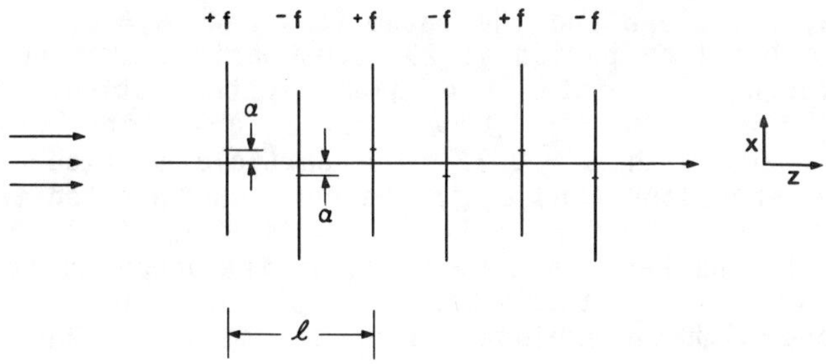

Fig. 2. Periodic thin-lens array used for simulation.

first traversed a focus lens whose center was displaced in the plus x-direction by an amount α. Next was a defocus lens displaced in the minus x-direction by the same amount, etc. The lenses had focal length f and were equally spaced with focusing period ℓ. This arrangement produces a net force on the beam in the plus x-direction. To determine the beam optical properties, we performed single-particle simulations. We learned that such alternating lens displacements produce a displaced neutral axis about which sinusoidal betatron motion takes place. The betatron frequency is independent of α; thus, in the spirit of the smooth approximation, there exists a constant average transverse force, and the betatron motion corresponds to that of a biased, simple harmonic oscillator.

In Fig. 3, the curve labeled $\alpha = 0$ shows the usual betatron motion about the z-axis. The particle started at the origin with a

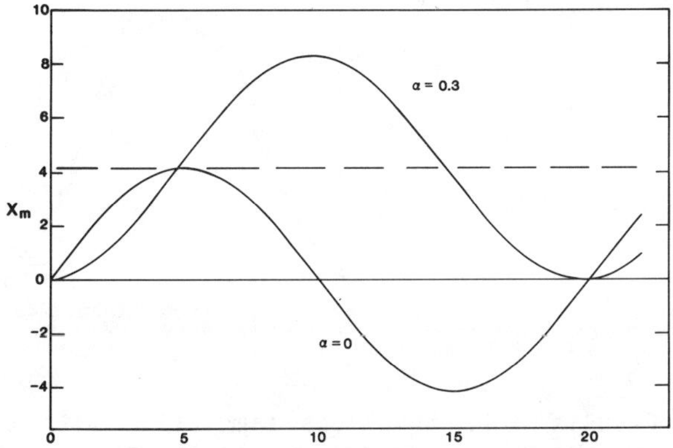

Fig. 3. Particle trajectories showing the effect of transverse lens displacements. With $\alpha = 0$, the betatron motion is centered on the z-axis. With $\alpha = 0.3$, the motion is centered about a displaced neutral axis.

positive initial slope and the parameters were $\alpha = 0$, $f = 16$, and $\ell = 10$. The betatron period is 20 focus periods and the curve is plotted through the points of maximum flutter motion. Also in Fig. 3, for $f = 16$ and $\ell = 10$, we show a curve for transversely displaced lenses with $\alpha = 0.3$. The particle started at the origin with zero slope, and again the curve was plotted through the points of maximum flutter. The curve is sinusoidal about a displaced axis and has the same betatron frequency as for $\alpha = 0$. In Fig. 4, we have plotted this same $\alpha = 0.3$ curve in more detail, along with an explanation of the motion. The top and bottom curves are drawn respectively through the maxima and the minima of the flutter motion. The average of these two curves is symmetric about the dashed neutral axis that is displaced from the z-axis by x_d. Note that the trajectory flutter is zero on the z-axis, increases to $\pm\alpha$ at $x = x_d$, and to $\pm 2\alpha$ at $x = 2x_d$. The left part of the figure explains these effects. In the smooth approximation, the transverse forces are proportional to the amplitude of the flutter motion. The <u>trajectory flutter</u> produces a restoring force proportional to the displacement from the z-axis. The <u>lens flutter</u> produces a constant (dipole) force of magnitude proportional to α, and these forces balance at $x = x_d$, producing the displaced neutral axis. The algebraic sum of forces from the two types of flutter then produces a restoring force that is linear relative to displacements from the <u>neutral</u> axis. Further,

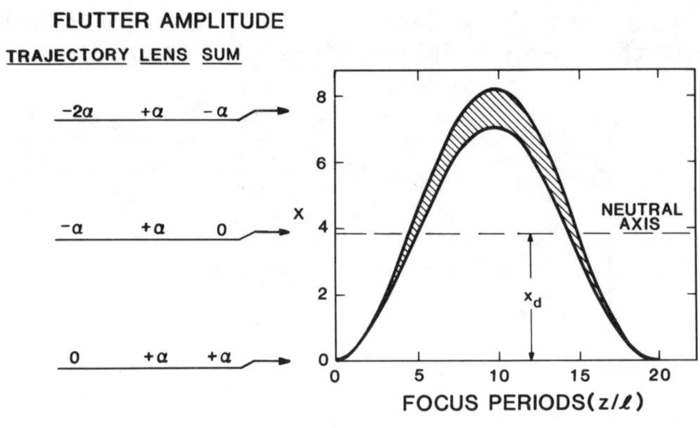

Fig. 4. Results of a thin-lens simulation with transverse alternating lens displacements of amplitude α. Initially, the particle was injected with zero slope at the origin. The upper and lower curves are plotted through the maxima and the minima of the trajectory flutter. The betatron motion is sinusoidal about the displaced neutral axis with a period independent of α. Amplitudes of the trajectory flutter, lens flutter, and their sum are shown on the left.

the betatron frequency is independent of α, so that the focusing properties are preserved.

To confirm that a displaced neutral axis exists, we simulated a particle starting at $x = x_d + \alpha$. For a proper choice of initial slope, the particle traveled along the lens system with flutter half-amplitude equal to α and with zero betatron amplitude, showing that forces from the lens and trajectory flutters exactly balance. It can be shown that the neutral axis displacement for this thin-lens system is $x_d = 8\alpha f/\ell$.

Next, we formulated a smooth-approximation calculation to describe these effects. The calculations correspond to an rf quadrupole focusing system that has successive lenses displaced transversely from the z-axis. We used a sine-wave displacement function of amplitude α and space period $\beta\lambda = 2\pi v/\omega$, where v is the particle velocity and ω is the rf quadrupole operating frequency. The appropriate differential equation is

$$M \frac{d^2x}{dt^2} + K(x - \alpha \sin \omega t) \sin \omega t = 0 \quad , \tag{1}$$

where K is the spring constant and M the particle mass. If $\alpha = 0$, Eq. (1) reduces to one form of the Mathieu equation used to describe periodic focusing. We assume a solution of the form

$$x = R \sin \Omega t + S \sin \omega t + x_d \quad , \tag{2}$$

where the first term represents the betatron motion, the second term the flutter, and x_d is the displacement of the neutral axis. If we take R>>S, and $\omega \gg \Omega$ (the smooth approximation), we obtain

$$S = \left(\frac{\omega_0}{\omega}\right)^2 (R \sin \Omega t + x_d) \quad , \quad \text{where} \quad \omega_0^2 = K/M \quad .$$

After substituting this into Eq. (2), Eq. (2) into Eq. (1), and then averaging the terms over one rf period, we find the equation satisfied if both

$$\Omega = \frac{\sqrt{2}}{2} \frac{\omega_0^2}{\omega} \quad , \quad \text{and} \quad x_d = \alpha\left(\frac{\omega}{\omega_0}\right)^2 = \frac{\sqrt{2}}{2} \alpha\left(\frac{\omega}{\Omega}\right) \quad . \tag{3}$$

First we see that the betatron frequency Ω is independent of α. Next we observe that x_d is equal to α multiplied by the factors $(\sqrt{2}/2)(\omega/\Omega)$. Because $\omega \gg \Omega$, then $x_d \gg \alpha$, which demonstrates that small transverse displacements are very effective in producing large displacements of the neutral axis. Although we have used the smooth approximation to obtain simple expressions for Ω and x_d, neither the beam manipulations discussed below nor the funneling scheme discussed later are necessarily limited to the region where the smooth approximation is valid.

Now we discuss the use of this deflection force to sidestep a beam from the z-axis to a displaced parallel axis. In doing this, we must use a method that does not induce additional betatron motion that would increase the beam emittance. One possible method is to start with a particle moving along the z-axis and, at the origin, turn on an initially constant value of α to induce a half-period betatron motion. When this excursion reaches its full amplitude, we then suddenly double the value of α to move the neutral axis to a point equal to the betatron amplitude. This procedure results in the displaced trajectory shown in Fig. 5, curve (a), where the curve is the mean of the flutter motion. After α has reached its final value of 0.6, this trajectory has a residual betatron amplitude of only 0.05% for the particular manner in which α was programmed. Alternatively, the particle could be injected along the displaced neutral axis with $x = 7.7$ and $\alpha = 0.6$. Then by reducing α in a programmed manner, the particle could be brought to the z-axis with zero slope. This sidestep maneuver forms one-half of the funneling scheme that in the next section will be extended so that two beams can be brought to a common z-axis. Also shown in Fig. 5 are curves (b) and (c) for which the motion was started with initial ± slopes. The amplitude of the resulting betatron motion about the displaced axis is equal to the case with $\alpha = 0$, showing that the sidestep maneuver produces no increase in the transverse emittance.

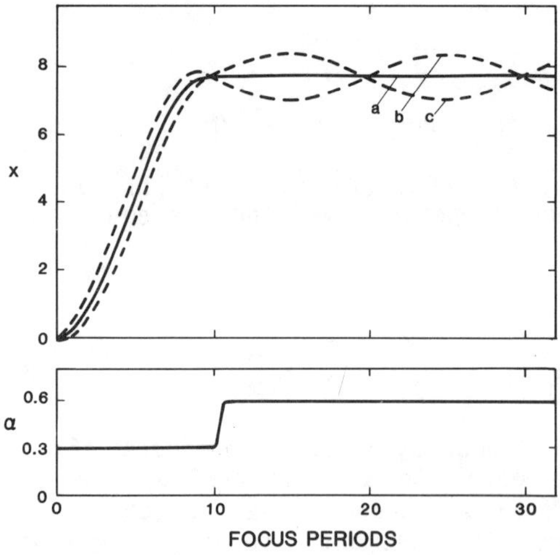

Fig. 5. Sidestep maneuver to displace a beam without increasing the transverse emittance. Particle (a) is injected with zero slope at the origin. Initially, a constant transverse force is applied through an alternating lens displacement of amplitude $\alpha = 0.3$. After one-half period of the betatron motion, the value of α is doubled to make the neutral axis coincide with the betatron maximum amplitude. The result is a trajectory displaced 7.7 units from the z-axis with negligible residual betatron amplitude. Particles (b) and (c) are injected with finite slope. They have the same amplitude as when $\alpha = 0$.

RFQ POTENTIAL FUNCTIONS

The conventional RFQ accelerator was proposed and developed in a somewhat unusual manner.[5,6] Instead of initially specifying the electrode geometry, a potential function was synthesized to give electric fields with the required beam-dynamical properties. Then, both the electric-field distributions and the isopotential shapes corresponding to the RFQ poles were derived from the potential function. We follow the same procedures in developing a new type of RFQ structure that has the beam-dynamical properties needed to funnel ion beams.

The standard RFQ accelerator potential function can be written in cylindrical coordinates (r, ψ, z) as

$$U = \frac{V}{2} [X \left(\frac{r}{a}\right)^2 \cos 2\psi + AI_0(kr) \cos kz] \sin (\omega t + \phi) \quad , \tag{4}$$

where V is the maximum potential difference between adjacent pole tips and $k = 2\pi/\beta\lambda$. Note that the I_0 Bessel function is even and symmetric in x and y. To provide a deflection force in the x-direction, we require I_0 to be replaced with a function that is odd in x and independent of y. Also, the potential function chosen must satisfy Laplace's equation. One example meeting these requirements is

$$U = \frac{V}{2} [C\left(\frac{x^2 - y^2}{a^2}\right) + D \sinh kx \cos kz] \sin (\omega t + \phi) \quad , \tag{5}$$

where the quadrupole focusing term is the same as in Eq. (4) but written in x-y coordinates. Another potential function is discussed in Appendix A.

The electric field components calculated from Eq. (5) are

$$E_x = - \frac{CV}{a^2} x - \frac{kDV}{2} \cosh kx \cos kz \ , \ E_y = \frac{CV}{a^2} y \ , \ E_z = \frac{kDV}{2} \sinh kx \sin kz \ ,$$

with each component multiplied by $\sin (\omega t + \phi)$. The second term in the expression for E_x produces a deflection force. The field averages taken over a βλ period, for a synchronous particle with x and y held constant, are

$$\bar{E}_x = - \frac{kDV}{4} \cosh kx \sin \phi \ , \quad \bar{E}_y = 0 \ , \quad \bar{E}_z = \frac{kDV}{4} \sinh kx \cos \phi \ .$$

Figure 6 compares the x-z plane pole-tip shape of the funneling RFQ with that of the conventional RFQ accelerator. The pole-tip shape of a funneling RFQ is discussed in Appendix B. The results of a numerical computation in the x-z plane are shown in Fig. 7. For the same parameters as Fig. 7, Fig. 8 shows the pole-tip shape in the x-y plane for three values of z. The transverse displacement of the pole tips in the x-direction with period βλ is the analogue of the thin-lens displacements of Fig. 2.

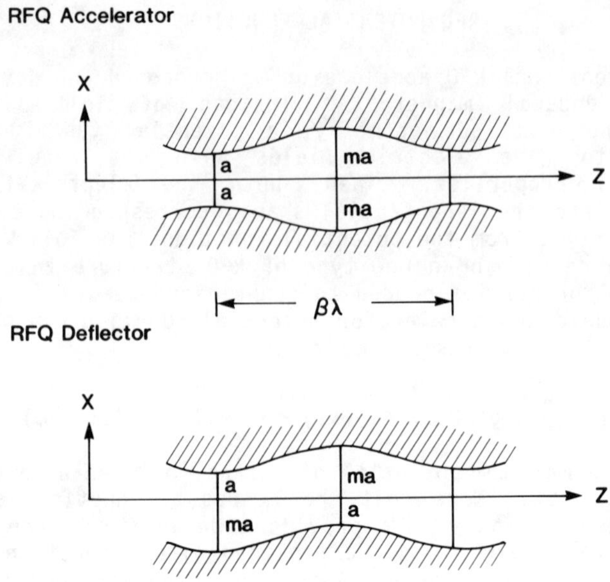

Fig. 6. Comparison of pole shapes in the x-z plane for an RFQ accelerator and for an RFQ deflector that can be used to funnel ion beams.

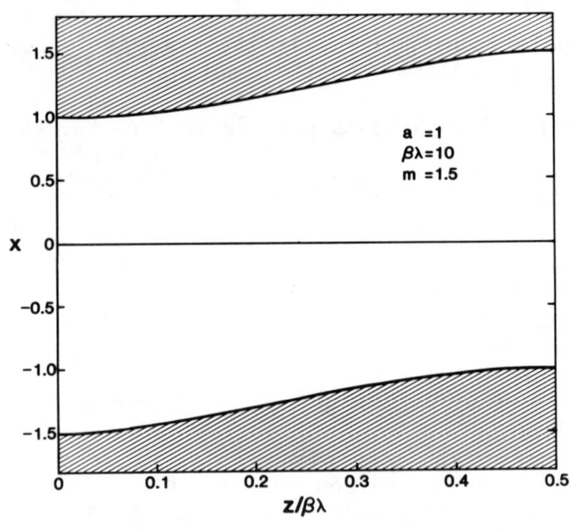

Fig. 7. Computed RFQ funnel pole-tip shape in the x-z plane.

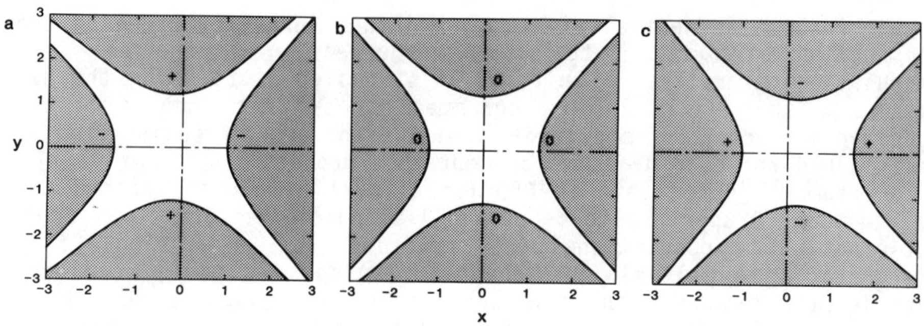

Fig. 8. An RFQ funnel pole-tip shape in the x-y plane at the beginning (a), middle (b), and end (c) of a unit cell of length $\beta\lambda/2$. The curves are from a numerical solution of Eq. (A-1) for a = 1, m = 1.5, and $\beta\lambda = 10$. The polarities shown correspond to a bunched beam with $\phi = -90°$. In addition to the usual quadrupole focusing forces, the beam will experience a deflecting force in the positive x-direction.

If we use the force corresponding to the above expression for E_x, the Mathieu equation becomes

$$\frac{d^2x}{dt^2} + \frac{qCV}{Ma^2}\left[x + \frac{ka^2D}{2C}\cosh kx \cos kz\right]\sin(\omega t + \phi) = 0 \quad . \qquad (6)$$

From a smooth-approximation calculation for a particle with synchronous velocity ($kz = \omega t$) and for $kx \ll 1$, so that $\cosh kx \simeq 1$, we obtain expressions for the betatron frequency Ω and for the displacement of the neutral axis x_d:

$$\Omega = \frac{\sqrt{2}}{2}\frac{\omega_o^2}{\omega} \ , \qquad x_d = \alpha\left(\frac{\omega}{\omega_o}\right)^2 = \frac{\sqrt{2}}{2}\,\alpha\left(\frac{\omega}{\Omega}\right) \ ,$$

where now

$$\omega_o^2 = \frac{qCV}{Ma^2} \ , \qquad \text{and} \qquad \alpha = -\frac{ka^2D}{2C}\sin\phi \quad .$$

These expressions show that the position of the neutral axis is proportional to $-\sin\phi$. If we introduce a bunched ion beam with $\phi = -90°$, x_d is positive and these ions can travel along their positively displaced neutral axis parallel to the z-axis. If another ion beam contains bunches with $\phi = +90°$, x_d is negative and these ions can travel along their negatively displaced neutral axis parallel to the z-axis. Therefore, we can inject two bunched beams having their microstructure bunches displaced by $\Delta z = \beta\lambda/2$ and have them travel along separate parallel trajectories displaced $\pm x_d$ from the z-axis. Both displaced beams will be transversely focused with the same strength about their respective neutral axes. We can then funnel by reducing the deflection parameter D to zero (by decreasing m, the pole modulation parameter, to unity). This enables us to bring the two beams to the z-axis in a programmed manner that induces no additional betatron amplitude. The transverse focusing is maintained for both beams throughout this process.

Beams injected into an RFQ funnel must be closely spaced to allow the RFQ aperture to be as small as possible. The beams could originate in a conventional RFQ structure that has a single rf resonator with multiple RFQ channels. Also, the two preceding accelerator channels could be designed with a sidestep near the output to bring their beams closer to the z-axis of the funnel. If the initial beams are not closely spaced, two RFQs based on Eq. (5) can be used to "pipe" the individual beams to the RFQ funnel. Alternatively, a periodic focusing system of static electric or magnetic quadrupoles (with their centers having programmed periodic transverse displacements) could be used, together with appropriate bunchers to prepare both beams for the final funneling operation. A septum magnet placed just before the funnel may also be necessary.

The expression for $\bar{E}_z$ is similar to that of a conventional rf linac for which the expression would be $E_0T \cos \phi$, where E_0 is the space average accelerating field and T is the transit-time factor. Because sinh kx is positive for positive x-displacement, such a beam will be bunched in a manner similar to a conventional linac beam with the added feature that the bunching forces are proportional to sinh kx. For $\phi = -90°$, the beam with positive x-displacement will experience both maximum positive-displacement forces and maximum bunching. Similarly, for $\phi = +90°$, the beam with negative x-displacement will experience both maximum negative-displacement forces and maximum bunching. The operating points for these two displaced beams are shown as asterisks in the Fig. 9 diagram, which portrays the full range of possible operating conditions. As the two beams approach the funneling vertex, the bunching forces are reduced to zero. If simulation studies show this to be a serious problem, it may be possible to devise an auxiliary method to maintain adequate bunching in the vertex region. One method to avoid possible debunching effects is to funnel rapidly. In the next section, we show an example where only one-fifth of a betatron period is required. Another possibility is to excite the poles of the funneling RFQ with both the fundamental and twice the fundamental frequency. The additional frequency would allow the superposition of the pole-tip modulation required to produce conventional bunching forces on the microstructure pulses, which now have $\beta\lambda$ spacing at the double frequency. The bunching force would have an I_0 Bessel function radial dependence and would be effective at all radii within the RFQ bore.

This type of structure can also serve as a two-beam accelerator. Two beams having bunches spaced $\beta\lambda/2$ can be injected along neutral axes displaced from the z-axis by $\pm x_d$. For example, if the RFQ is designed so that the beams at $+x_d$ has $\phi = -45°$, and the beam at $-x_d$ has $\phi = +135°$, the expression for

$$\bar{E}_x = -\frac{kDV}{4}\cosh kx \sin \Phi$$

$$\bar{E}_z = \frac{kDV}{4}\sinh kx \cos \Phi$$

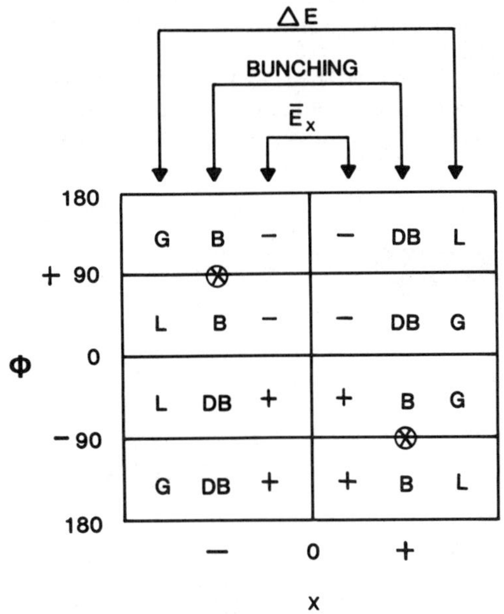

Fig. 9. Diagram showing the full range of operating conditions that are possible with a funneling RFQ. The energy gain G or loss L, the bunching B or debunching DB, and the sign of the transverse deflecting field $\bar{E}_x$ are shown as a function of the beam displacement x, and the synchronous phase ϕ.

$\bar{E}_x$ and $\bar{E}_z$ show that both beams will be accelerated, in addition to being bunched, focused, and held in displaced trajectories.

FUNNELING EXAMPLE

If we neglect space charge, funneling trajectories can be obtained by integrating both the transverse equation, Eq. (6), and the equation obtained using the longitudinal field E_z:

$$\frac{d^2z}{dt^2} - \frac{qVkD}{2M}\sinh kx \sin kz \sin(\omega t + \phi) = 0 \quad .$$

An unoptimized set of parameters was chosen to demonstrate the general properties. Protons of 2-MeV energy entered a 100-MHz RFQ funnel with an aperture parameter $a/\sqrt{C}$ = 33 mm. The peak voltage between RFQ poles was 264 kV, corresponding to a peak surface field of about 10 MV/m, and a betatron phase advance of 15° in the focus period of $\beta\lambda$ = 196 mm. The results are shown in Fig. 10(a). Particles are injected along $\pm x_d$ with $\phi = \mp 90°$, and each particle follows its neutral axis until a deflection force is introduced at $z/(\beta\lambda)$ = 2.75 to deflect them toward the z-axis. This force is generated by programming the parameter α, which changes the deflection parameter D. The variation of α is shown in Fig. 10(b). To funnel rapidly, large changes are made in α to first deflect both particles toward the z-axis, and then at $z/(\beta\lambda) = 4.75$, α is reversed in sign to provide both particles with a deflection force away from the z-axis. Finally, at $z/(\beta\lambda) = 7.75$, α is made equal to zero at the funneling vertex where both particles have reached the z-axis with zero slope. The forces required to funnel extend from $z/(\beta\lambda) = 2.75$ to 7.75, or 5.0 focus periods. The betatron period is 24 focus periods, so that the funneling operation required 21% of a period. In this example, there was no change in the longitudinal velocity because synchronous particles of $\phi = \pm 90°$ were injected so that $\bar{E}_z \equiv 0$. Figure 10(a) also shows the RFQ pole-tip shape in the x-z plane. The cross-hatched poles are plotted with the same vertical scale as

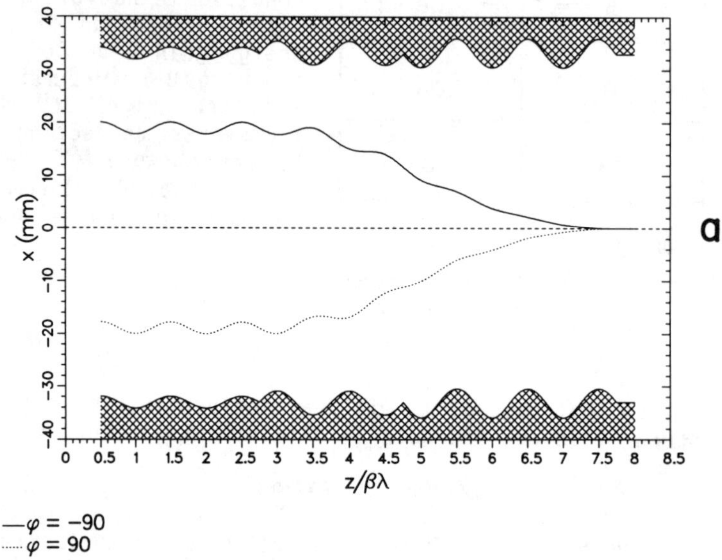

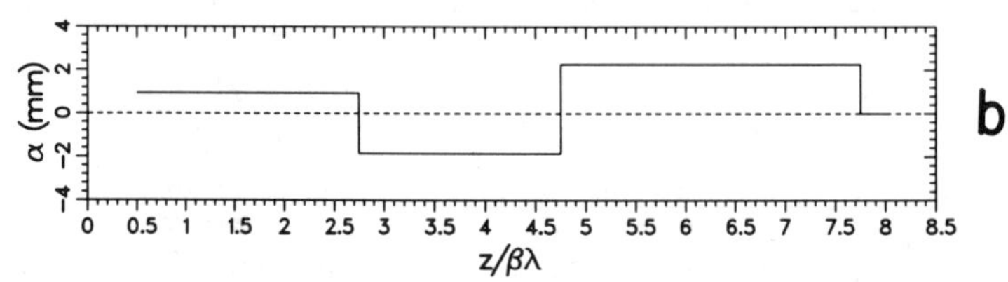

Fig. 10. RFQ funnel calculated trajectories. In Part (a), particles are injected on either side of the z-axis with $\phi = \pm 90°$. They are then deflected so that they reach the axis with zero slope to form a collinear beam of interlaced microstructure pulses. Part (b) shows the programming of the lens displacement parameter α necessary to provide the appropriate deflections. During the process, all particles will be subject to uninterrupted transverse focusing toward their respective central trajectories.

that of the particle trajectories. At two places, the poles have a sharp cusp-like shape that would not be acceptable from the standpoint of rf breakdown criteria. These cusps are an artifact of using square-wave programming for α. This situation can easily be avoided by appropriate smoothing.

Another set of calculated trajectories is shown in Fig. 11(a), with the programming of x for φ = -90° shown in Fig. 11(b). To illustrate the effectiveness of the transverse focusing throughout the funneling process, we initiated all the trajectories one-half betatron wavelength ahead of the funneling vertex. First, trajectories with zero betatron amplitude were calculated for beams on either side of the z-axis. Then for positive x, two trajectories

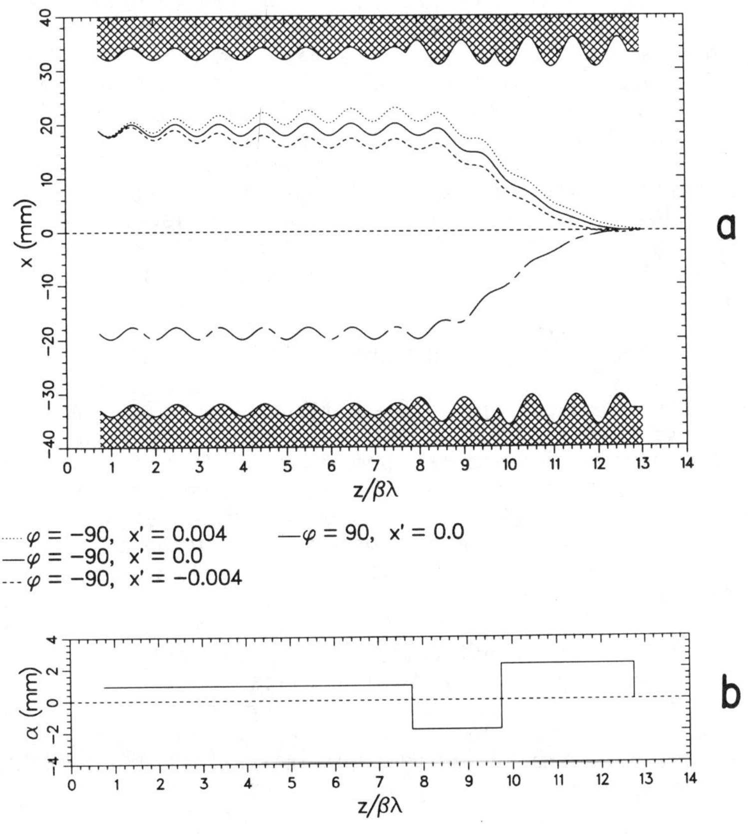

Fig. 11. Demonstration of the transverse focusing properties of the RFQ funnel. In Part (a), trajectories were calculated for particles injected one-half betatron wavelength ahead of the funneling vertex. All trajectories reach the z-axis at the funneling vertex, which shows the continuous nature of the transverse focusing with negligible effects from field nonlinearities. In Part (b) the programming of α is shown for φ = -90°.

with finite betatron amplitude were calculated to observe the effect of transverse focusing on a beam of non zero emittance. Both trajectories reach the z-axis at the funneling vertex. This result is similar to Fig. 5, except that electric field nonlinearities whose magnitudes depend upon kx could perturb the betatron period and produce emittance growth. For the trajectories of Fig. 11, kx varied from 0.63 to zero; and the absence of nonlinearity effects suggests that RFQ funneling can be accomplished at appreciably large values of kx. This means that beams can be combined in a funnel of reasonably large aperture without requiring especially high ion velocities or low RFQ frequencies.

The funneled beam has unusual transverse phase-space characteristics. At the funnel output in a specific transverse plane, successive microstructure pulses have alternating phase-space ellipse parameters. In this situation one must use time-varying optical elements to match the beam. One possibility is to use a radial matching section[6] similar to that used at the input of an RFQ accelerator. Adding such a radial matching section to the RFQ funnel output will make all pulses have identical ellipse parameters and will produce an output beam equally divergent in both transverse planes. Then, a solenoid to produce a convergent beam will allow it to be matched to an RFQ accelerator operating with frequency 2f. The 2f RFQ accelerator would consist of an input radial matching section, a rebuncher, and a final accelerator section that could feed another funnel or an Alvarez linac.

CONCLUSION

To funnel high-current, low-emittance ion beams at low velocities, we have proposed a new type of RFQ structure and have investigated its beam-dynamical properties. This RFQ provides strong electric periodic focusing to the ion beams, and at the same time provides the transverse deflection forces necessary to merge two bunched beams into a single beam with interlaced microstructure pulses. The mechanical and rf design for the RFQ funnel should be greatly facilitated by the widespread active development of the RFQ accelerator. It should be feasible to apply existing technology in the areas of pole-tip construction, resonator design, and rf power. Also, RFQ accelerator beam-dynamics design and analysis techniques such as the RFQUIK and PARMTEQ programs can be adapted. These practical considerations, as well as the intrinsic properties of the RFQ funnel, should make it an attractive candidate for increasing beam brightness.

APPENDIX A

ALTERNATIVE FUNNELING POTENTIAL FUNCTION

A second RFQ potential function that may be useful for funneling is

$$U = \frac{V}{2}\left[G\,\frac{2xy}{a^2} + H \sinh kx \cos kz\right] \sin(\omega t + \phi)$$

The second term is identical to that of Eq. (5), but the quadrupole focusing term has been rotated 45° about the z-axis. In this case, periodic modulation of the poles in the y-direction produces deflection forces in the x-direction. This is a new geometry that could be advantageous in certain applications. Using the above potential function allows replacement of one pair of poles by an isopotential conductor in the y-z plane, which permits acceleration, bunching, and focusing of a single displaced beam with only two poles, plus the isopotential plane.

APPENDIX B

FUNNELING RFQ POLE-TIP SHAPES

To find the shape of the isopotential surfaces of the RFQ poles, we use Eq. (5) and set U(max) = V/2 to obtain

$$\frac{C}{a^2}(x^2 - y^2) + D \sinh kx \cos kz = 1 \quad . \tag{A-1}$$

Let $y = 0$ and set $x = a$ for $z = 0$, and $x = ma$ for $z = \beta\lambda/2$. We then obtain expressions for the coefficients C and D:

$$C = 1 - D \sinh ka \quad , \qquad D = \frac{m^2 - 1}{m^2 \sinh ka + \sinh mka} \quad .$$

Numerical solutions of Eq. (A-1) were used to calculate the electrode shapes shown in Figs. 7 and 8. Although these figures for illustration use a large value of the deflection parameter m, often much smaller values will be sufficient to produce the desired deflection force. For example, if the phase advance per focusing period is 36°, $ka = 0.1$, and $\phi = -90°$, the neutral axis is displaced by a/2 if $m = 1.14$.

REFERENCES

1. D. E. Young, Proc. Heavy Ion Fusion Workshop, Brookhaven National Laboratory report, BNL-50769, 17, (1977); R. Kustom et al., ibid., 94; D. A. Swenson, Proc. Heavy Ion Fusion Workshop, Lawrence Berkeley Laboratory report, LBL-10301 and Stanford Linear Accelerator Laboratory report, SLAC-PUB-2575, 239, (1979).

2. K. Bongardt, Proc. 1984 Linac Conf., Gesellschaft für Schwerionenforschung, Darmstadt report GSI-84-11 (September 1984), 389.

3. F. W. Guy and T. P. Wangler, and F. W. Guy, these proceedings.

4. E. A. Knapp and R. A. Jameson, Accelerator Technology Program, January-June 1981, p.46, Los Alamos National Laboratory report LA-9320-PR (May 1982).

5. I. M. Kapchinskii and V. A. Teplyakov, Prib. Tekh. Eksp. No. 2, 19-22 (1970).

6. K. R. Crandall, R. H. Stokes, and T. P. Wangler, Proc. 10th Linear Accelerator Conf., Brookhaven National Laboratory report, BNL-51134, 205 (1979).

THE FOM-MEQALAC EXPERIMENT

R.W. Thomae, F. Siebenlist, P.W. van Amersfoort, F.G. Schonewille, E.H.A. Granneman*,
FOM-Institute for Atomic and Molecular Physics,
Kruislaan 407, 1098 SJ Amsterdam, The Netherlands

H. Klein, A. Schempp, T. Weis,
Institut für Angewandte Physik,
University Frankfurt, Robert-Mayer-Strasse 2-4, 6000 Frankfurt, FRG

ABSTRACT

The FOM-MEQALAC (Multiple Electrostatic Quadrupole Array Linear Accelerator) system in which four He^+ ion beams of 2.2 mA are accelerated from 40 to 120 keV is discussed. Some fundamental considerations concerning the design of MEQALAC systems are presented. We report measurements of currents, emittances and energies of injected 10 mA beams, performed behind the ion source, matching section, Low Energy Beam Transport (LEBT) section and the accelerator. Furthermore, parameters for plasma diagnostics and ion implantation accelerators are given.

INTRODUCTION

Energetic particle beams with a high brightness are required for many applications, such as diagnostics of a fusion plasma or surface treatment of metals. The space charge induced blow-up of the ion beam sets severe limits to the maximum current which can be accelerated. With the MEQALAC concept this problem is circumvented by dividing the beam into a large number of parallel beamlets, which are accelerated simultaneously.

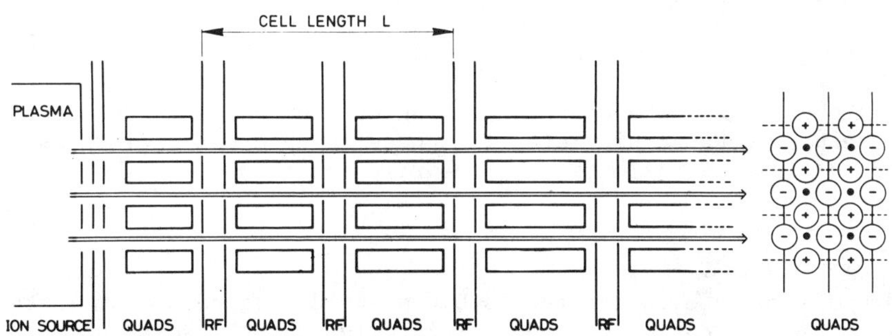

Fig. 1. The MEQALAC concept. A large number of beams is injected into an RF gap / electrostatic quadrupole structure and accelerated to a high energy. The special arrangement of the quadrupole elements makes it possible to stack many beams within a small area.

*Present address: ASM Europe, Bilthoven, The Netherlands

A schematic of the MEQALAC concept is shown in Figure 1. A number of beamlets is injected into a multichannel RF structure. RF voltages are applied to a (usually) large number of acceleration gaps. In each gap a fixed amount of energy is gained by the particles. Electrostatic quadrupole lenses are placed in between the acceleration gaps to oppose a space charge induced blow-up of the beams. Furthermore, they counteract the radial defocusing effect of the acceleration field. By increasing the number of acceleration channels the total current is increased. Note that the individual beamlets are effectively shielded from one another by the quadrupole elements. By stacking the channels as shown in the right side of Figure 1 a dense packing of the beamlets is obtained. This system was invented by Maschke[1]. Because limited experimental information is available on acceleration of space charge loaded beams, we decided to do a proof of principle experiment which is meant to acquire the necessary expertise and scaling laws. Subsequently, an accelerator with direct applications in a major branch of physics or technology will be built[2].

DESIGN CONSIDERATIONS

Maschke[3] and Reiser[4] showed that an approximate solution can be derived for the space charge limited current of DC beams in periodic focusing systems, if the modulation in the beam envelope is small. It is derived with Reiser's so-called smooth approximation that the maximum current I_{MAX} which can be transported through a periodic focusing channel scales with the particle velocity β, the channel radius a_Q, cell length L, so-called (zero current) phase advance per cell μ_o, particle mass number A and particle charge state Z as[4]

$$I_{MAX} \sim \frac{A}{Z} \beta^3 \mu_o^2 \left(\frac{a_Q}{L}\right)^2 \qquad (1)$$

The cell length L is defined as

$$\tfrac{1}{2} L = \ell + g \qquad (2)$$

where ℓ is the length of a quadrupole singlet lens and g is the distance in between successive singlets. The phase advance per cell μ_o is described in first approximation by the relation[4]:

$$\mu_o \approx \frac{ZU}{A} \frac{\ell^2}{a_Q^2 \beta^2} (1 + g/\ell) \qquad (3)$$

The quadrupole voltage is denoted as U. To avoid dangerous beam instabilities[5] μ_o should not exceed 90°, preferably $\mu_o \lesssim 60°$. Relation (1) shows that a small cell length L is favourable for increasing the maximum current density. The influence of fringe fields in a quadrupole singlet lens, however, increases strongly with decreasing ℓ/a_Q. Using relation (2) with a minimum gap distance g, we find that L and a_Q cannot be chosen independently. A working rule is

$$a_Q \lessdot 0.1\ L \qquad (4)$$

Equation (1) and (4) imply that the maximum obtainable value of I_{MAX} is independent of the channel radius a_Q. The same result is obtained for bunched beams. Therefore the channel radius a_Q has to be chosen as small as possible to increase the current density of a MEQALAC system. On the other hand, the emittance of the ion source has to be smaller than the acceptance α of the quadrupole channel. This parameter scales as[4]

$$\alpha \sim a_Q^2/L \tag{5}$$

This implies that a first limit to the minimum channel radius a_Q is set by the ion source. Furthermore, to keep μ_o at a constant value, the quadrupole voltage U has to be kept constant for decreasing a_Q and constant ratio a_Q/ℓ (Equation (3)). The breakdown voltage, however, decreases with decreasing a_Q. Finally, the alignment and mechanical stability of the quadrupole array are critical at small values of a_Q. Taking these considerations into account we believe that $a_Q = 2 - 3$ mm is a practical lower limit for the channel aperture radius. The corresponding lower limit for the cell length is $L = 20 - 30$ mm. Note that this distance is occupied by two gaps and by two quadrupole singlet lenses.

The large capacitive loading in the MEQALAC resonator allows a representation of the resonator by a so-called lumped-circuit model. The resonance frequency f_o is related to the inductance L_R and capacitance C_R of the resonator cavity,

$$f_o = (4\pi^2 L_R C_R)^{-\frac{1}{2}} \tag{6}$$

L_R is proportional to the square of the cavity diameter D and inversely proportional to the total cavity length L_C:

$$L_R \sim D^2/L_C \tag{7}$$

The resonator capacitance C_R can be described in terms of structure dimensions:

$$C_R \sim \frac{F}{d}\frac{L_C}{L} \tag{8}$$

where F and d are the gap area and width, respectively. Moreover, $2L_C/L$ equals the number of gaps. For a fixed resonance frequency f_o, the so-called parallel resonance resistance R_{po} scales as[13],

$$R_{po} \sim C_R^{-\frac{3}{4}} L_R^{\frac{3}{4}} L_C^{5/2} L^{-2} \tag{9}$$

At a given resonator frequency, a large cavity diameter D is favourable for having a large R_{po} value, which is in turn favourable for minimizing the RF power losses in the cavity.

EXPERIMENTAL SET-UP

A schematic picture of the FOM-MEQALAC experiment is shown in Figure 2.

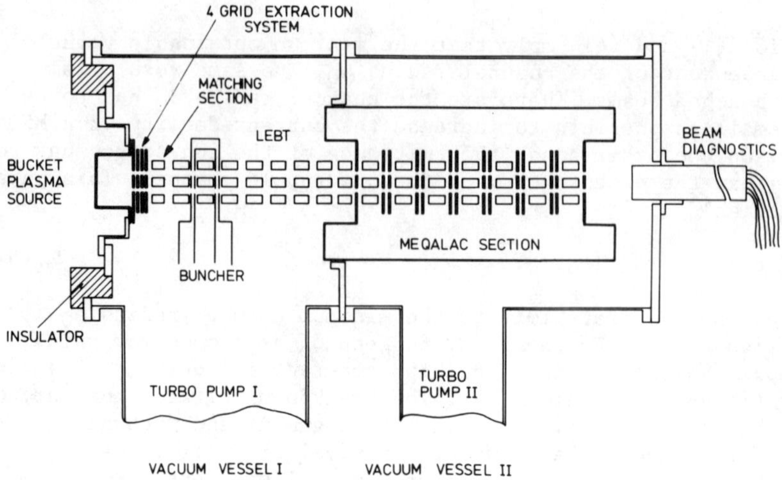

Fig. 2. The experimental set-up. For details, see text.

From a bucket type ion source four He^+ ion beams are extracted. A Low Energy Beam Transport (LEBT) section transports the ions from the high pressure ion source region to the low pressure RF acceleration section[6,7,8]. Each channel of the LEBT section consists of a series of 34 electrostatic quadrupole singlet lenses, the first of which is located 18.7 mm behind the last grid of the extraction system. All LEBT singlets are held at the same potential except for the first five

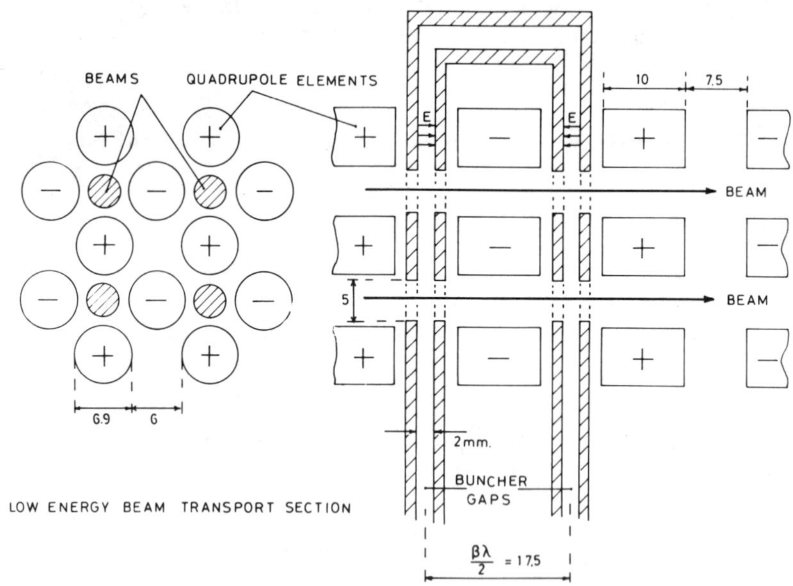

Fig. 3. Characteristic dimensions (in mm) of the LEBT section and of the two-gap buncher.

quads which can be excited independently. Halfway down the LEBT section the DC beams are bunched by means of a two-gap buncher. The characteristic dimensions of the LEBT section and of the two-gap buncher are given in Figure 3. A quadrupole voltage of ± 2.4 kV is required to achieve a zero-current phase advance per cell of 60° for the 40 keV He^+ ion beams. For a typical beam emittance of ~20 π mm mrad and a maximum filling factor of $F = a^2/a_Q^2 = 70\%$, a = maximum beam radius, a_Q = channel radius, the theoretical space charge limited current is 18 mA[6].

The 40 MHz two-gap buncher consists of a capacitively shorted λ/4 resonator. The gaps are at 17.5 mm distance. In between the two gaps one quadrupole singlet is situated, in order to continue the periodicity of the transport section[9]. The drift length from the buncher up to the end of the LEBT section is 35 cm.

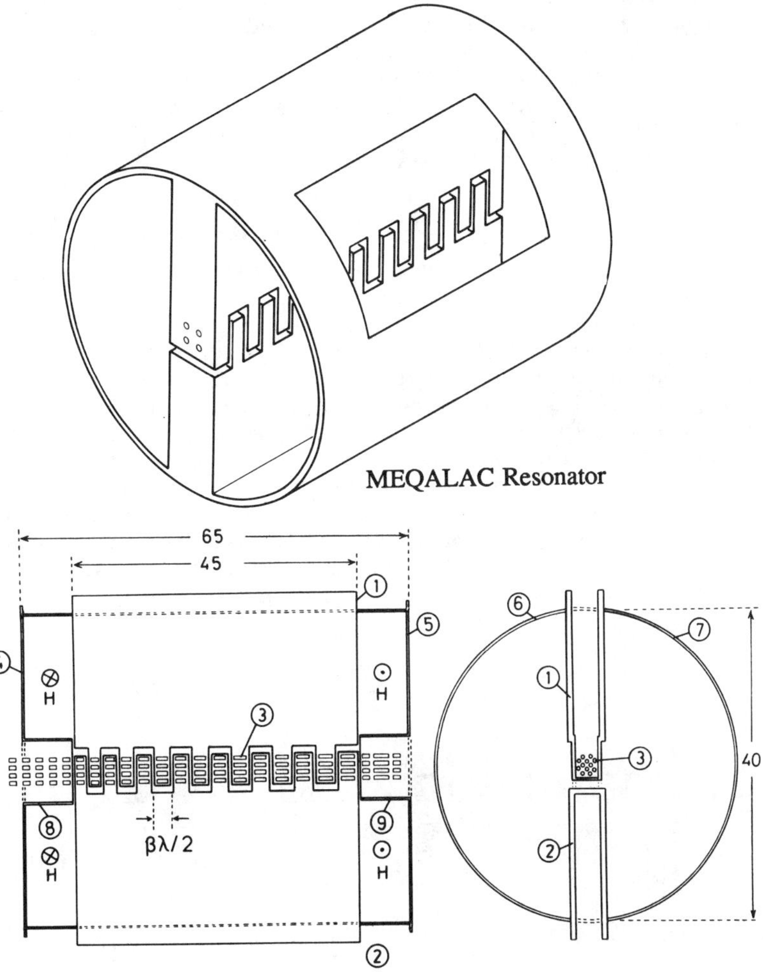

Fig. 4. The 40 MHz MEQALAC acceleration structure. All dimensions are in cm.

Figure 4 shows a schematic picture of the modified Interdigital H resonator, in which the particle bunches are accelerated at a resonance frequency of 40 MHz. Two hollow rectangular boxes (1,2) which carry fingerstructures are mounted on opposite sides of a cylindrical cavity. The RF magnetic field which is flowing around the boxes provides a strong coupling of the cells. The quadrupole elements (3) are placed inside the fingers, where they are shielded effectively from the RF field in the gaps.

The length of the elements increases proportionally to the anticipated particle velocity. Hence the focusing forces remain constant for a given quadrupole voltage $U_{q,m}$. The space charge limited current for one channel is calculated to be of the order of 3 mA for a quadrupole voltage $U_{q,m} = \pm$ 2.4 kV and an RF input power of 700 W, which corresponds to a synchronous phase of -38°. For this calculation it is assumed that the injected current is matched perfectly in the radial as well as in the longitudinal plane.

Low level perturbation ball measurements show that the RF field distribution is flat within 7%. The resonance frequency is measured to be 40 MHz, the R_{po} value 16 MΩ and the unloaded quality factor 1800. Further parameters of the accelerator are given in Table 1, Column 1.

MEASUREMENTS

The matching of space charge loaded beams to the acceptance of a periodic focusing system is of crucial importance for stable transport and acceleration. The beam properties are determined at several places along the beam line, which are indicated in Figure 5 by A, B and C. In the entrance plane of the first lens of the matching section, indicated by A in Figure 5, first the current and the emittances of one of the beams is measured. The beam envelope and its derivative are deduced from the emittance diagram. The periodic solution for the focusing channel and the required voltage setting of the matching quadrupoles are calculated with a computer program, which employes the so-called kV equations[5,10]. In this program the potential field of the quadrupoles is calculated from the Laplace equation. The average r^2-dependence of the field as a function of z is used.

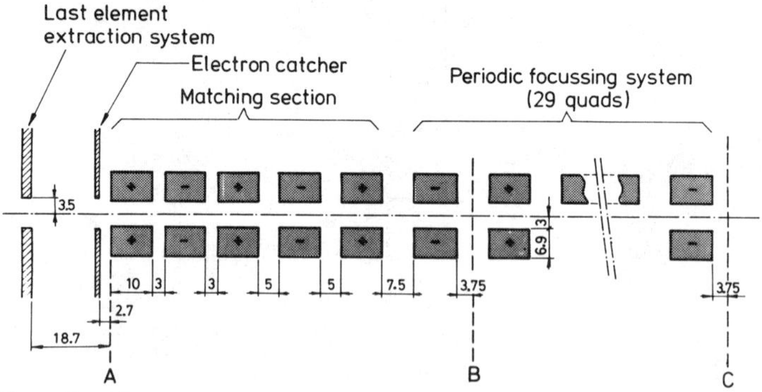

Fig. 5. Characteristic dimensions of the LEBT section. The three planes in which the emittance diagrams are measured are indicated by A, B and C. All dimensions are in mm.

The next step is the installation of the matching section together with the first quadrupole lens of the periodic channel. Behind this section, indicated by B in Figure 5, the emittance diagrams are measured in the xz and yz planes with the calculated voltages applied to the matching quadrupole elements. Finally the whole LEBT section is installed. The emittance diagrams are measured in the plane indicated by C in Figure 5, i.e. behind the last quadrupole element. The emittance diagrams of a 10 mA beamlet as measured at the three indicated positions, are shown in Figure 6.

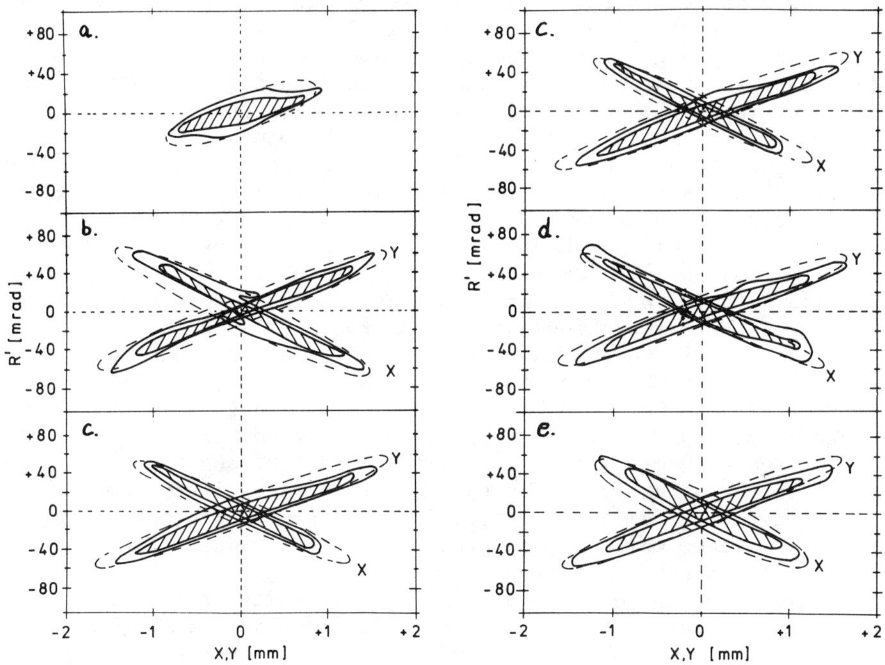

Fig. 6. Emittance diagrams. The outer solid lines indicate the contours where the intensity is 10% of the maximum value. The solid lines enclosing the shaded area represent the 1/e contours. The dashed lines give the RMS ellipse as deduced from the measured intensity distribution.

a: Emittance diagram of a 10 mA beam at the entrance plane (figure 5,A) of the first quadrupole. The RMS values are: envelope radii $X = Y =$ 0.86 mm, envelope derivatives $X' = Y' =$ 25.9 mrad, $\varepsilon = 17.3\ \pi$ mm mrad.

b: Emittance diagram of the beam x and y planes, measured behind the sixth quadrupole (figure 5,B). The x and y diagrams are not fully symmetric because the actual location of the entrance slit of the emittance measuring device is 2 mm behind its front plate. This grounded front plate is placed at the centre of the gap separating two successive quadrupoles. $X = 1.45$ mm, $Y = 1.64$ mm, $X' = -64.2$ mrad, $Y' = 58.5$ mrad, $\varepsilon_x = 26.8\ \pi$ mm mrad, $\varepsilon_y = 24.3\ \pi$ mm mrad.

c: Emittance diagram of the beam x and y planes, measured behind the last quadrupole of the LEBT section (figure 5,C). $X = 1.25$ mm, $Y = 1.68$ mm, $X' = -51.3$ mrad, $Y' = 56$ mrad, $\varepsilon_x = 17.3\ \pi$ mm mrad, $\varepsilon_y = 24.9\ \pi$ mm mrad.
d: Emittance diagram of the beam x and y planes, measured behind the LEBT section; the buncher is excited with $P_B = 60$ W, $U_B = \pm\ 1.75$ kV, bunch length at the end of the LEBT section ~ 9 ns.
e: The buncher is excited with $P_B = 200$ W, $U_B = \pm\ 3.2$ kV, bunch length at the end of the LEBT section ~ 6 ns.

The measured and calculated values agree within 15%. The current measurements show that 15% of the current is lost inside the matching section, while practically no beam loss is observed in the transport section. These measurements are carried out with a quadrupole voltage of ± 2.4 kV ($\mu_o = 60°$) in the LEBT section. The matching quadrupole voltages range from 1.3 kV to 3.9 kV. When the buncher is excited the time averaged transported current is found to decrease gradually from 8.5 mA to 7.5 mA when the buncher input power increases from zero to 400 W. This power corresponds to a total buncher peak voltage of $U_B = \pm\ 4.5$ kV. The minimum bunch length of 6 ns is measured at $P_B = 200$ W, $U_B = \pm\ 3.2$ kV. The emittance diagrams measured while the buncher is excited with 60 and 200 W are shown in Figure 6. Note that the dimensions and the orientations of the emittance ellipses do not change as a function of RF power. In contrast to our measurements the calculations done with the kV-transport program show a considerable increase in beam envelope. The action of the buncher is modelled in this time-independent program via the assumption of a current linearly increasing with distance between buncher and exit of the LEBT section. The discrepancy is partly explained by the fact that the emittance measuring device integrates over 500 ms, i.e. it measures the time averaged emittance diagram, while in the program only the middle (the "peak value") of the bunch is traced.

A large fraction of this 8.5 mA beamlet, which is ~ 3 times larger than the space charge limited current of an accelerator channel, is expected to get lost, when it is injected. Therefore the acceleration experiments were carried out with one beam line only. The exit energy of the bunches is determined by means of a time of flight method. In Figure 7 the accelerated current I_{cup} is shown as a function of the resonator input power P. The buncher is not excited. The measurements are done for different values of the accelerator quadrupole voltage $U_{q,m}$. The LEBT quadrupole voltage $U_{q,\ell}$ is kept constant at $U_{q,\ell} = 2.4$ kV ($\mu_o = 60°$). Therefore, a certain radial mismatch has to be taken into account when $U_{q,m} \neq U_{q,\ell}$. The accelerated current I_{cup} in Figure 7 is observed to increase with increasing RF power P for small values of P. In this region we expect that the current is "longitudinally limited". Above a certain value of P the radial defocusing space charge forces are expected to become dominant and to determine the limiting current. This implies that the current is constant for a further increase of P, as is seen in Figure 7. It is also seen that the maximum current which is accelerated drops dramatically for quadrupole voltages $U_{q,m} > 4.0$ kV. A possible explanation for this reduction is on one hand the increasing radial mismatch and on the other beam instabilities[5], which are predicted for quadrupole voltages which correlate with phase advances

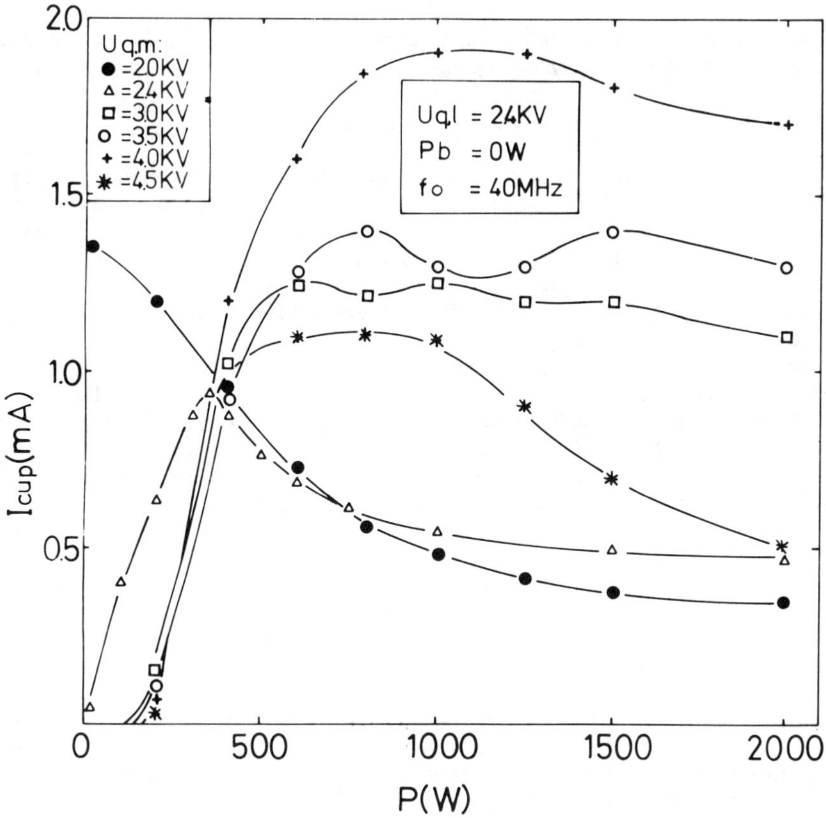

Fig. 7. The accelerated current I_{cup} as a function of the resonator input power P for different quadrupole voltages $U_{q,m}$. The injected current is ~ 8.5 mA/beamlet. The buncher is not excited.

$\mu_o \geq 90°$. Note that a quadrupole voltage $U_{q,m}$ = 4.0 kV corresponds already to a phase advance per cell of 110°. For quadrupole voltages $U_{q,m} \leq 2.0$ kV and a small input power, non-accelerated particles are observed to travel through the resonator. For $P \geq 300$ W all particle bunches are accelerated to the design energy of ~ 120 keV.

The accelerated current increases by 10 - 30% when the buncher is excited. A maximum current of 2.2 mA/beamlet is accelerated when the buncher and resonator cavities are excited with 100 W and 1 kW, respectively. In this case the quadrupole voltage in the accelerator is $U_{q,m}$ = 4 kV. First calculations with a multiparticle code show that the major particle losses take place in the first cells of the accelerator and are caused by the space charge of the bunches. It was found from the simulations that the bunch peak current is of the order of 35 mA.

Future experiments with different beam currents (1 mA $\leq I \leq$ 15 mA) and other diagnostic equipment, such as a magnetic energy analyser and a fast 50 Ω Faraday cup will allow for a more quantitative discussion of the results. Furthermore, emittance measurements will be carried out behind the accelerator.

MEQALAC DESIGNS

A list with the characteristic parameters of the MEQALAC designs is given in Table 1. The designs are means for different applications, such as plasma diagnostics (column 2) or modification of metal or semiconductor surfaces (column 3 and 4). The design of a 6 MeV, 100 mA Li^- MEQALAC system had already been presented at the 1984 LINAC conference[8]. The parameters of the present MEQALAC experiment are given in column 1 for comparison. All designs are a first approach.

In the accelerator given in column 2 a 190 mA D^- ion current is accelerated from 80 keV to 1 MeV. This system can be used for diagnostics of a fusion plasma[11]. The same system can be used for the acceleration of a 95 mA H^- current from 40 keV to 500 keV.

The accelerators in column 3 and 4 are primarily designed for implantation of nitrogen ions into metal surfaces[12]. They can be used for the acceleration of a 110 mA N_2^+ ion current from 80 keV to 2 MeV (3), or for the acceleration of a 102 mA N^+ ion current from 40 keV to 1 MeV (5). These accelerators can also be used, for instance, for implantation of P^+ of B^+ ions into a semiconductor surface. The required beam current is then in the μA range[12], so that only one channel of the accelerator has to be used.

Table 1. Characteristic properties of a MEQALAC design for plasma diagnostics (2) and modification of a metal/semiconductor surface (3 and 4). The parameters of the present experiment are shown in column 1. The subscripts L and T denote the longitudinal and transverse dimensions, respectively. The acceleration efficiency is defined as the fraction of the total RF power which is converted into particle power.

Parameter	1	2	3	4	Dim.
Particle	He^+	D^-	N_2^+	N^+	-
Injection energy	40	80	80	40	keV
Exit energy	120	1000	2000	1000	keV
RF frequency	40	80	27	25	MHZ
Synchronous phase	-38	-30	-20	-20	°
Gap electric field amplitude	2.6	12.0	14.2	12.0	MV/m
Width RF gaps	0.2	0.4	0.5	0.4	cm
Number of gaps	20	23	33	24	-
Number of channels	4	25	36	64	-
Overall beam dimensions	4	35	35	65	cm^2
Length resonator	65	150	200	170	cm
Diameter resonator	40	40	100	80	cm
Quality factor	1800	2500	3700	2800	-
Parallel resonance resistance R_{po}	16	28	110	38	MΩ
$R_{po,eff}$	8.1	17	79	27	MΩ
$\beta\lambda/2$ first cell	1.75	1.95	1.60	1.81	cm
$\beta\lambda/2$ last cell	2.80	6.10	6.50	7.40	cm
Quad spacing/length; g/l	0.75	0.95	1.30	0.81	-
Channel radius	0.30	0.30	0.25	0.30	cm
Quadrupole voltage ± U	2.6	6.3	6.7	3.3	kV
Zero current μ_{oT}	60	60	60	60	°
Zero current μ_{oL}	19.8	27.6	30.5	35.8	°
Depressed μ_T	24.0	24.0	24.0	24.0	°
Depressed μ_L	7.9	11.0	12.2	14.3	°
Channel acceptance α_T	108 π	97 π	95 π	104 π	mm mrad
Channel acceptance α_L	270 π	112 π	100 π	130 π	mm mrad
I_T time averaged	2.9	7.7	3.1	1.6	mA
I_L time averaged	3.1	7.6	3.2	2.3	mA
Total current	11.6	190	110	102	mA
Acceleration efficiency	54	78	83	74	%

REFERENCES

1. A.W. Maschke, Brookhaven Natl.Lab., report BNL-51209 (1979).
2. J. Kistemaker, H.J. Hopman and E.H.A. Granneman; The MEQALAC project, FOM rep.nr. 52.580/1 (1981).
3. A.W. Maschke, Brookhaven Natl.Lab., report BNL-51022 (1979).
4. M. Reiser, Part.Acc. 8, 167 (1978); J.Appl.Phys. 52, 555 (1981).
5. I. Hofmann, L.J. Laslett, L. Smith and I. Haber, Part.Acc. Vol.13 (1983) p.145.
6. P.W. van Amersfoort, E.H.A. Granneman, J. Kistemaker, F. Siebenlist, R.W. Thomae, H. Klein, A. Schempp; The FOM-MEQALAC project, status report November 1982, FOM rep.nr. 55.120 (1982).
7. F. Siebenlist, R.W. Thomae, P.W. van Amersfoort, E.H.A. Granneman, H. Klein, A. Schempp, T. Weis; Proc.of the 1984 Lin.Acc.Conf., Seeheim, FRG, GSI-Report 84-11, p.363.
8. R.W. Thomae, F. Siebenlist, P.W. van Amersfoort, E.H.A. Granneman, H. Klein, A. Schempp, T. Weis; Proc.of the 1984 Lin.Acc.Conf., Seeheim, FRG, GSI-Report 84-11, p.118.
9. R.W. Thomae, P.W. van Amersfoort, F. Siebenlist, F.G. Schonewille, E.H.A. Granneman, J. Kistemaker, H. Klein, A. Schempp, T. Weis; The FOM-MEQALAC project, status report December 1983, FOM rep.nr. 57.379 (1983).
10. I.M. Kapchinsky, V.V. Vladimirski, Proc.Int.Conf.on High Energy Acc., CERN, Geneva, 1959, p.274.
11. D.E. Post, L.R. Grisham and R.J. Fonck, Physica Scripta T3, 135 (1983).
12. Proc.5th Int.Conf.on Ion Implantation: Equipment and Techniques; eds. H.H. Andersen and S.T. Picraux, Nucl.Inst.Meth.in Phys.Rev. B6 1,2 (1985).
13. J. Pottier, IEEE Trans.Nucl.Sci., NS-16, No.3, 372 (1969).

HIGH-CURRENT SIMULATION CODES

I. Haber
Naval Research Laboratory, Washington, DC. 20375

ABSTRACT

The suitability of using P.I.C. codes to simulate intense ion beams is discussed by presenting examples of their use. The numerical methods employed in these codes, as well as their limitations, are explained and some examples are presented to illustrate possible consequences of these limitations.

INTRODUCTION

Recent experimental and theoretical progress in transporting high current ion beams with very low emittances has substantially increased interest in understanding the detailed behavior of accelerators which produce very high beam intensities. At these intensities, space charge forces become increasingly dominant so that the beam evolution can depend not only on averaged quantities such as beam current and emittance, but can also be strongly influenced by details of the six-dimensional distribution function. Measurement of detailed information about the distribution function of a very low emittance beam is usually very difficult. In addition, the complexity of individual particle orbits in an alternate gradient transport system which includes strong space charge forces can render a detailed theoretical treatment exceedingly difficult. Numerical simulations have therefore emerged as an important tool in understanding the dynamics of high intensity particle beams as well as an aid in both the interpretation of experimental results and in the design of intense beam accelerators.

The simulations discussed here employ numerical techniques, originally developed for the general simulation of plasma dynamics, which have been modified to address the behavior of intense charged particle beams. The intent is to discuss, primarily by example, when these techniques are most appropriate to accelerator simulations. Since accelerator simulation codes have been incorporating increasingly sophisticated models of space charge forces, at the same time as the plasma physics codes are incorporating more accelerator physics, the distinction between accelerator codes and plasma physics codes has become increasingly blurred. Sufficient differences still remain, however, so that the two approaches can indeed differ substantially in emphasis, and more importantly, in the results obtained from the simulations. A section on the capabilities of these codes, particularly as evidenced by examples of work from the Heavy Ion Fusion program is presented. The next section is a short description of the numerical techniques and some of the codes employed. A discussion follows of some of the numerical pitfalls which have been encountered in these examples.

BACKGROUND

Since the emittance of a beam is very difficult to reduce, the requirement for a very low emittance at the output of an accelerator system usually translates into a requirement that emittance growth be limited at all stages of the beam lifetime. The use of a heavy ion beam to ignite a thermonuclear pellet is a good example of such a system - where substantial engineering and cost advantages can be obtained by maximizing beam intensity. Particularly in the linear induction accelerator approach to ignition, moreover, the relatively simple geometry has made it possible to examine an important portion of the relevant physics without consideration of a three-dimensional system with complex boundaries. The large ratio of beam length to radius, for example, makes the separate treatment of the longitudinal and transverse dynamics a good approximation to the actual accelerator.

In the transverse dimension, a matched periodic transport system with an assumed equilibrium distribution function can be characterized by two free parameters. One of these, which is usually the phase advance per cell, is a feature of the transport system without space charge. The other, which is a measure of the beam intensity, can be one of several equivalent and related quantities such as the depressed phase advance, the ratio, I/ε, of the beam current to emittance, or the ratio of plasma frequency to betatron frequency. When the beam is sufficiently intense so that the space charge forces become comparable to the external focusing forces, the plasma frequency also becomes comparable to the betatron frequency and the phase advance per cell is depressed well below the low current value. This is the same parameter regime where space charge collective behavior can significantly influence the gross beam dynamics and use of the Vlasov equation, along with its associated nonlinearities, is necessary to follow the evolution of the beam space charge forces.

A large number of simulations have been performed to investigate the behavior of such a space charge dominated beam. A linear perturbation theory has shown[1] that a K-V equilibrium is unstable in this parameter regime. Since the perturbations assumed in this analysis identically preserve rms emittance, however, the linear theory is not a suitable vehicle to examine the growth of emittance. In fact even though the theory predicts instability over a wide range of phase advances, simulations have found, and to a major degree these findings have been verified experimentally,[2,3] that for transport systems with low current phase advances well below 90°, instabilities do not cause emittance growth. An idealized transport system, therefore, can transport currents of arbitrarily high intensity without degradation.

A method which has been used[4,5] to examine a transport system over a large range of currents, is to simulate the behavior at the center of a long beam being longitudinally compressed by gradually increasing the current in the simulation, which is in the beam frame. Figure 1 shows the evolution of the x and y rms emittances of a transport system with an initial K-V distribution and enough current to depress the phase advance from 60° to 12°. The current is held constant for 50 periods, linearly increased by a factor of ten over the next 100 periods, and then held constant for 50 more periods. The negligible increase in rms emittance is in contrast to the behavior, shown in Fig. 2, of a 90° transport system depressed to 15° and with the current increase similarly programmed.

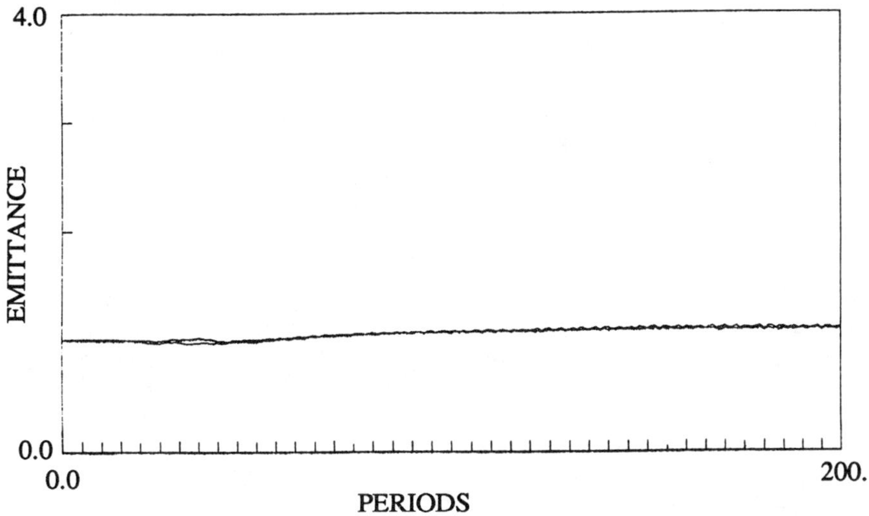

Fig. 1. Evolution of the x and y rms emittances during transport of a beam with an initial K-V distribution and enough current to depress the 60° phase advance to 12° . The current is held constant for 50 periods, increased tenfold over the next 100 and finally held constant at the higher value for the last 50.

Because of the analytic tractability of the Kapchinskij-Vladimirskij (K-V) distribution, early simulations were largely confined to the study of K-V systems. Studies of non K-V distributions[4] have shown behavior not markedly different from the K-V system, though the behavior of such non K-V systems can be complicated by initial conditions which are far from equilibrium. At very low phase advances, however, the general form of the equilibrium is better understood,[6] and it is relatively easy to construct initial distributions which are close to equilibrium. This occurs because the beam space charge almost cancels the average focusing force and the particles, which follow almost straight line phase space orbits within the beam, are turned around by the large potential change which appears in a thin region near the beam edge. Though nonuniformities in the beam spatial distribution can substantially affect the emittance, the details of the velocity distribution become increasingly less important and affect only the details of the beam edge region. This is illustrated by examining the evolution of a semi-Gaussian, or thermal, distribution which is uniform in space and

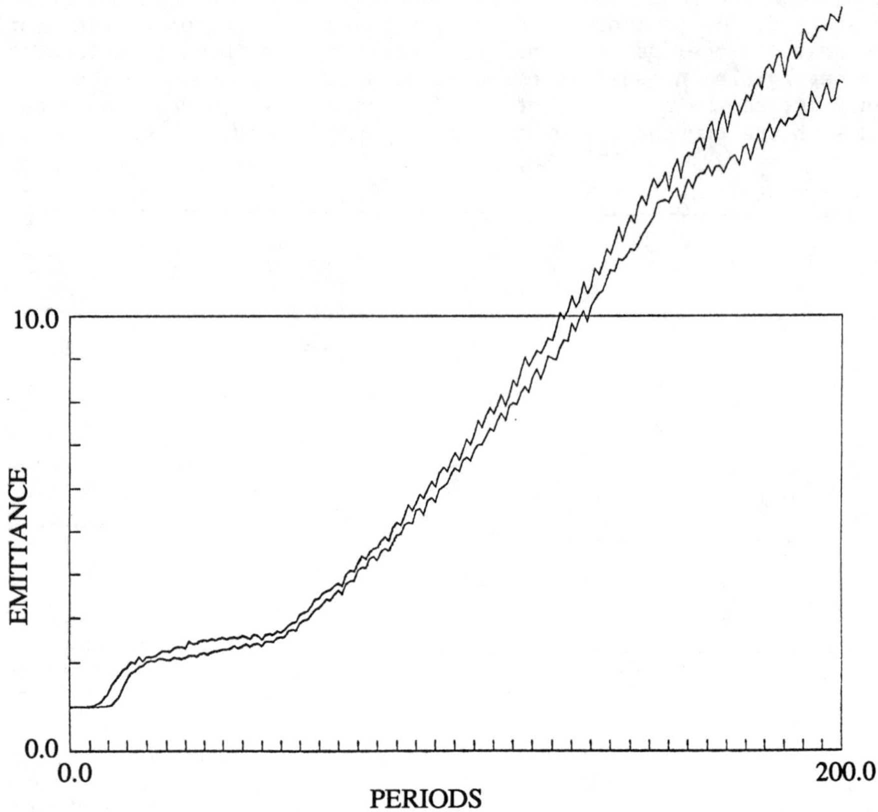

Fig. 2. Contrasting behavior of the x and y rms emittance in a $90^{\circ}/15^{\circ}$ system with the current programmed as in Fig. 1.

has a Gaussian velocity distribution independent of position. Figure 3 shows four views of the phase space of such a distribution for a 60° transport system with a depressed phase advance that would be 12° for a K-V distribution with the some rms properties. Figure 4 shows this system after it has been propagated through 25 periods. Except for a rounding of the beam edges in configuration space, very little change is seen. Since a distribution which is uniform in space and Gaussian in velocity is a relatively realistic approximation to what is experimentally attainable, this is further evidence that very low phase advances should be possible to achieve in practice.

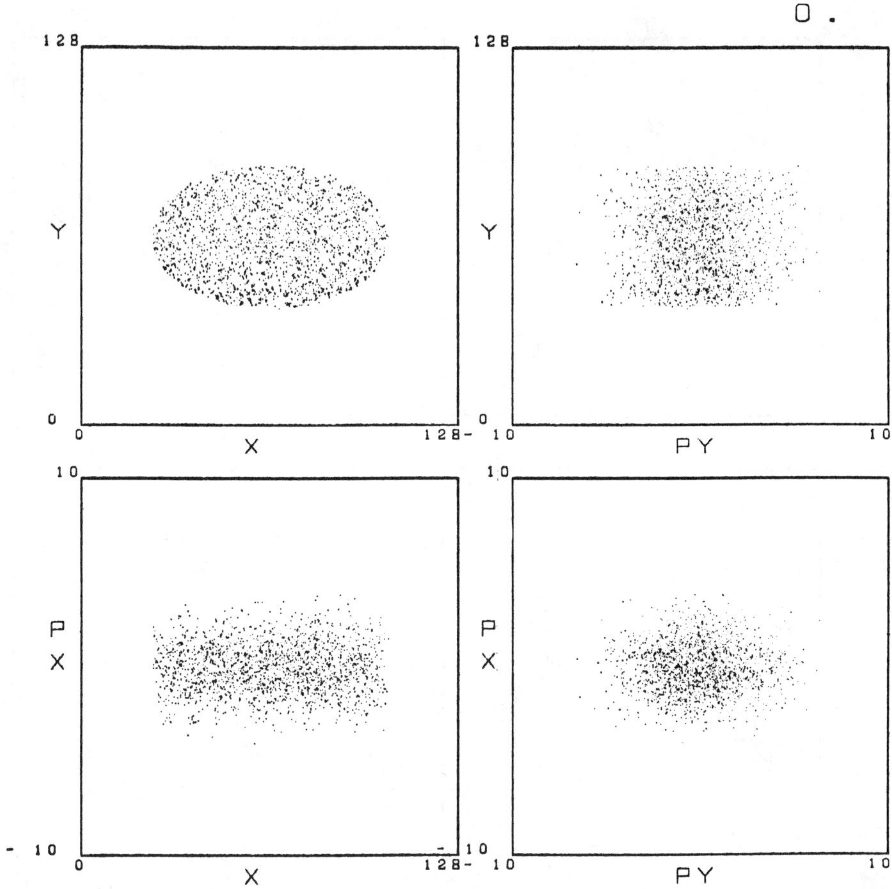

Fig. 3. Four phase space views of an initial distribution uniform in configuration space and with a Gaussian velocity distribution independent of position.

These two simulation examples were chosen to illustrate some of the ways in which simulations have increased confidence in the achievability of very low emittance transport. They are, however, only a small part of the many published and unpublished examples too numerous to cite here. These cover a variety of topics including a survey of work on instabilities,[7] and examples of work oriented towards specific accelerator design issues such as the consequences of conducting boundaries[8] and external field nonlinearities[9,10] on centered and off-centered beams, and the effect of space charge on multi-turn injection into a storage ring.[11]

A large body of simulation results also exists in connection with the study of the longitudinal dynamics of beams long compared with their diameter. These range from study of the fundamental properties of waves on a beam with the very low energy spreads that

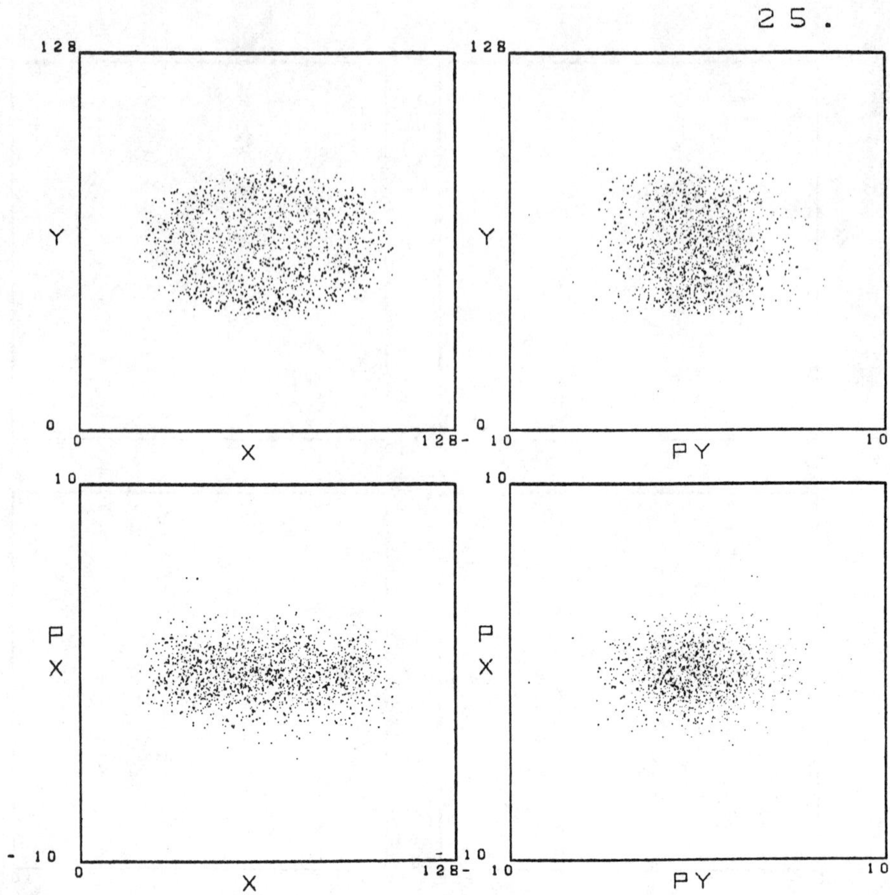

Fig. 4. Four phase space views of the distribution in Fig. 3, after it has traversed 25 periods, showing little change except for a rounding of the edges in configuration space.

characterize Heavy Ion Fusion beams[12], to detailed studies of the beam dynamics associated with the actual parameters expected in ignition systems. Some examples include the parametric study of longitudinal instabilities[13,14] and bunching[14-18] with many of the complications expected in driver systems.

The limited number of examples cited thus far have in common their use of a class of algorithm developed for the simulation of plasmas. There are many more examples of published and unpublished results using these codes as well as a large number which use different kinds of numerical algorithms. The next section discusses numerical techniques which are common to these plasma codes. In the last section, some examples are presented of numerical pitfalls whose avoidance makes the use of these techniques an appropriate way to simulate accelerator physics in the intense beam parameter regime.

NUMERICAL ALGORITHMS

Since thermonuclear, laboratory and ionospheric plasmas are generally many Debye lengths in extent, and a large number of particles in a Debye sphere is a fundamental characteristic of collisionless plasmas, plasma simulations generally require large numbers of particles. The algorithms which have been developed to simulate these plasmas therefore are designed to follow efficiently large numbers of particles in their self-consistent electromagnetic fields. These computer codes generally employ the particle-in-cell method of advancing the dynamical system. By accumulating the particle quantities on a grid, and then solving Maxwell's equations on the grid to obtain the forces which are used to advance the particles, the operation count generally scales linearly with the number of particles. This is in contrast to any algorithm which sums up the forces on each particle from all the others and therefore scales quadratically with the number of particles.

In addition to the large number of published papers on the numerics of plasma simulation codes, several excellent volumes have been published[19-23] which contain detailed analysis of the use of these codes and the numerical methods on which they are based. Because of the interest here in use of these codes for low emittance ion beams, only a brief discussion of the "standard" electrostatic algorithm and its use for this type of simulation will be presented.

The choice of algorithm for any class of problems is often a complicated tradeoff between speed and accuracy that includes a large number of other factors such as the reliability, ease of use, and availability of various computer codes. Existing plasma codes have several advantages. Many have been used for a long time and are therefore well debugged and they have often been highly optimized for a variety of computer hardware configurations. Though these codes are usually large in the sense that they contain many lines of code, because of the simplicity of the algorithms, they tend to be highly modular and therefore easy to modify even when not as well documented as desirable. They are often big because they contain a large number of diagnostic and physics options which can prove useful in unanticipated ways if the physics to be modelled proves to be more complex than originally expected. For example, if the necessity to follow electron trajectories suddenly seems important, the ability to simulate multiple species is a common feature of these codes.

On the other hand, there are some limitations in using these codes for accelerator problems. For example, if it is necessary to follow the particle orbits to high accuracy, this may require a major rewriting of the code and may change its characteristic behavior sufficiently so that the numerical behavior must be reinvestigated. Perhaps more importantly, in many accelerators, the characteristic time and length scales are determined by external forces and not by the characteristics of the beam plasma, so that not many particles are needed to simulate the system adequately. A numerical algorithm which was designed under the assumption that integration of particle orbits dominates the running time, may then prove less than efficient.

The use of an existing computer program also has its disadvantages, particularly if it is very large. Modification of a large code to do a simple problem can sometimes prove to be more time consuming than writing a new code for that problem.

By describing the principles behind the simple electrostatic particle code, describing some examples of those codes which have been used to do accelerator simulations, and then discussing the numerical requirements which make the intense beam regime possibly hard to simulate, it is hoped to aid the reader in making an informed judgement on these tradeoffs.

The accuracy-speed tradeoff in plasma codes is usually resolved in favor of relatively simple low order temporal differencing with a high degree of symmetry. Especially in view of the truncation errors in solving the field equations on a grid, higher order accuracy in timestepping is not usually a good tradeoff. In the electrostatic limit of interest here, the time advancement of the numerical system is just a staggered leapfrog. This staggered leapfrog is time-centered and reversible, properties termed "virtuous" by Langdon, and it is in fact possible to show that this sort of algorithm has several desirable properties, including symplectic.[24]

The system of equations to be solved is a combination of the equations of motion of the individual particles in their self and external fields and the electrostatic limit of the field equations:

$$\frac{dx_i}{dt} = v_i$$

$$\frac{dv_i}{dt} = \frac{q_i}{m_i} E(x_i) + \frac{F_{ext}}{m_i}$$

$$\rho = \sum_i q_i \; x_i$$

$$\nabla \cdot E = 4\pi\rho$$

where x_i and v_i are a position and velocity component of the i'th particle, E is the electric field, ρ is the charge density, q_i and m_i are the charge and mass of the i'th particle and F_{ext} is the external force applied to the particles.

The time advancement is shown in Fig. 5. The particle positions, the charge density and the external forces are defined at the center of a time step while the velocities are defined a half step between. The bottom half of Fig. 5 shows a typical particle orbit which is composed of straight line segments and is advanced in time according to:

$$x_i(t+\Delta t) = x_i(t) + v_i(t+\Delta t/2)\,\Delta t$$

$$v_i(t+3\Delta t/2) = v_i(t+\Delta t/2) + \frac{q\Delta t}{m} E(t+\Delta t) + \frac{\Delta t}{m} F_{ext}(t+\Delta t)$$

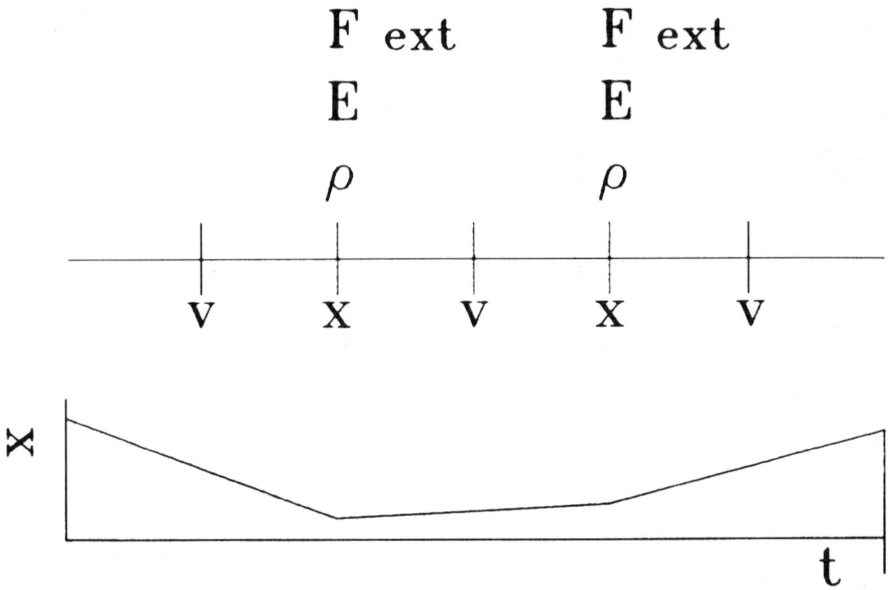

Fig. 5. Time diagram of the "standard" leapfrog electrostatic simulation algorithm, showing in the top half, the definition of the grid quantities and external forces at the same set of times as the particle positions and the particle velocities at staggered times. Also drawn, in the bottom half, is a plot of a typical particle orbit in x-t space, showing the similarity of the algorithm to a thin lens focusing systems.

One way of describing the result of a staggered leapfrog is that the particle orbits are moved, to machine roundoff, through a series of thin lenses. Since both the external forces and the space charge forces can be represented as thin lenses, the code is accurate to the extent that this thin lens approximation is adequate for the timestep used. That is, the particle displacement during a timestep must be small compared with the local field gradient. It is of course always possible to divide the focusing and space charge forces to obtain enough thin lenses to approximate them as accurately as desired. However, this may not be consistent with the computer resources available for the particular problem at hand.

Though a large class of existing codes employ the staggered leapfrog described above to advance the particle orbits in time, a variety of methods are utilized to solve for the self-consistent fields from the particle distribution. Some variety of methods also exists for accumulating the particle quantities, which are defined at the locations of each particle, onto the charge density matrix which is defined only at the grid points. A similar process is necessary for calculating the force on each particle which is also defined only on the grid points. The most common such process is probably area weighting, which attributes the particle and field quantities according to the overlap between a rectangle drawn with each grid point as the center and the same size rectangle drawn with a particle at the center.

The variety of field solving methods is more extensive. As an example the SHIFT series of computer codes, used by the author and also at LBL[1-3,8-10,12,13,17], employs a fast Fourier transform (FFT) Poisson solver. FFT's are a fast method for solving the field equations and have some advantages in their ability to tailor the particle-particle interaction by smoothing the fields. However, they are generally appropriate only to a rectangular mesh. The SHIFT codes have therefore been implemented in two-dimensional rectangular (XY) geometry and a one dimensional (Z) geometry, which is appropriate for beams long compared with their diameter.

The SCOP codes used by Hofmann and coworkers[1,7,14-16], primarily at GSI and the Max Planck Institute at Garching, use finite differencing to solve for the fields and is therefore not limited to rectangular coordinates. The SCOP codes have been used extensively in XY rectangular and RZ cylindrical versions.

A numerical approach which uses finite elements is contained in the TRACK codes used by Prior at Rutherford[11]. This method is suitable for complex geometries including complicated boundary conditions, at the expense of some complication in the field solver.

The MASK code of Drobot at SAIC, used by Bisognano, et al.[18] represents a slightly different approach because the code is a very general purpose implementation of the algorithms discussed above. By using a combination FFT tridiagonal solve field solver, XY, Rθ, and RZ geometries are available as well as subroutines to treat various types of physics such as emitting surfaces, appropriate to several classes of problems. In fact, since the code is fully electromagnetic and relativistic, only a small part of its capabilities are necessary for ion beam problems. And though the large size of the code which is measured in tens of thousand of lines could be a problem, the successful use of MASK, and other codes of this size, to solve a large number of problems over a wide range of parameters by a large number of investigators working on a variety of computer hardware configurations certainly suggests that any problems caused by using a code of this size can be overcome.

Both the MASK and SHIFT codes have the ability to accommodate complicated conducting boundaries using a capacitance matrix[25,26]. This approach exploits the linearity of the electric field equation. The potential at each of a number of boundary points is found due to a charge at each of the other boundary points. A matrix is then constructed which relates the potential at each boundary point to the charges at all the others. This matrix is then inverted to obtain a capacitance matrix which when multiplied by the potential at each boundary point will give the charge distribution at the boundary points required to produce that potential. In particular, if the potential chosen is just the negative of the potential which resulted from the interior charges, a zero potential at the boundary is obtained when the boundary charges are added to the original set and the fields solved a second time. Though its most common use is to produce a grounded potential this method can be used to produce any potential distribution on the boundary. Furthermore, because linearity of the fields was the only feature of the numerical method assumed, the capacitance matrix is suitable for use with almost any type of field solver.

This section has discussed the operation of a typical plasma electrostatic particle code and mentioned a small number of examples of the implementation of this class of algorithms. The last section will attempt to discuss some of the requirements for successful modeling of beam transport problems, including some examples of where difficulties in the numerical simulation may occur.

NUMERICAL LIMITATIONS IN BEAM SIMULATIONS

The numerical solution of a system of partial differential equations in space and time usually requires timesteps which are short compared with the shortest characteristic time, and spatial resolution finer than the gradient scale lengths characteristic of the physical system to be modeled. In addition, when a numerical system is used to model the Vlasov equation by following the evolution of a large number of macroparticles, it is additionally required that the number of particles be sufficient that statistical fluctuations, arising from the unphysically small number of particles in the numerical system, do not affect the physics.

The requirement that the timestepping be rapid compared with the plasma frequency is usually superseded, when simulating beam propagation, by the requirement that the timesteps be short compared with the time variation associated with the external focusing as viewed in the beam frame. This translates into a requirement for several timesteps per focusing period or per betatron period in the presence of a continuous focusing system.

The usual requirement for spatial resolution to be of order the plasma Debye length is also superseded. In fact, the most conservative approach to determining required spatial resolution would require gridding fine compared with the density gradient at the beam edge. However, in practice this is usually too conservative since the errors encountered by a particle at the beam edge appear to cancel as a particle moves into and out of the edge region.

This type of "better than might be expected behavior" has its dangers. Though numerical tests on a beam at the system center would lead to the conclusion that a certain system is not resolution limited, use of the same numerical parameters to simulate an off-centered beam in the presence of images in electrostatic focusing electrodes[8] can be used to generate an interesting counterexample.

Figure 6 is a plot of the x and y rms emittances of this off-centered beam. The beam center is offset by about a quarter of its radius at a 45° angle. Though the numerical system has 128 x 128 cells, after allowing for the electrodes and the correct ratio of beam diameter to electrode spacing desired, only about 32 cells remain with which to resolve the beam. A characteristic of this off-centered beam is the presence of several different frequencies which are comparable in magnitude including the betatron frequency, depressed betatron frequency, the actual frequency of the coherent beam oscillation in the presence of the image forces and the plasma frequency. The lack of cancellation of edge errors because the beam is moving can therefore have a measurable effect.

Figure 7 is the same physical system but this time run on a 256 x 256 mesh. In this case the emittance, plot which should look the same as Fig. 6 because the same initial distribution of particles was used, shows only about half the emittance growth as in the 128 x 128 run. A simple but expensive way to gain further insight into whether the emittance growth is entirely numerical, which could be inferred from the factor of two difference in growth accompanying a factor of two difference in resolution , is to run the 512 x 512 system which is shown in Fig. 8. In this case though there are some small differences in the microstructure of the emittance plots, the 256 x 256 and 512 x 512 curves fall almost on top of each other.

The possibility of inadequate resolution at the edge of a beam where the density gradients are very sharp, is only one area where inadequate care in the numerics can give spurious answers. Another set of possible problems can occur in the general area of particle-particle collisions.

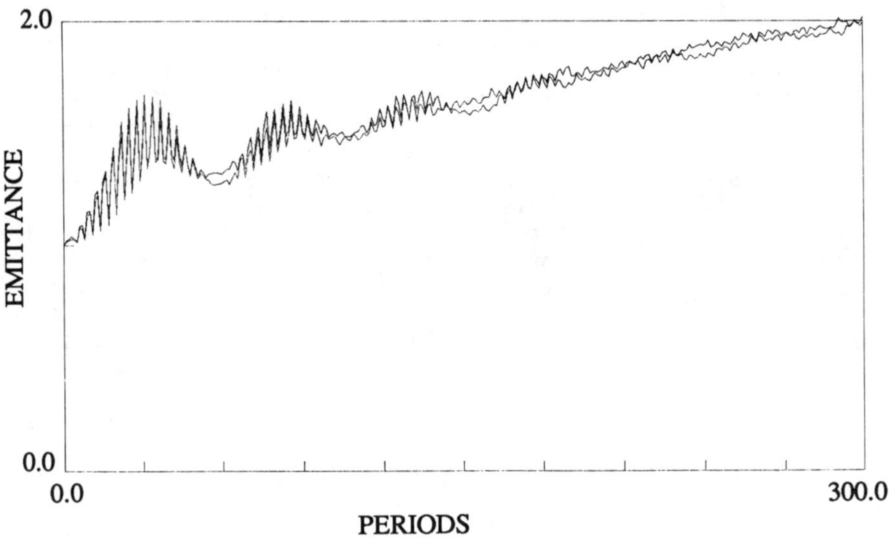

Fig. 6. Plot of the x and y rms emittances as an off-centered beam is propagated down a focusing channel with electrostatic quadrupole lenses. The numerical system used a 128 x 128 grid.

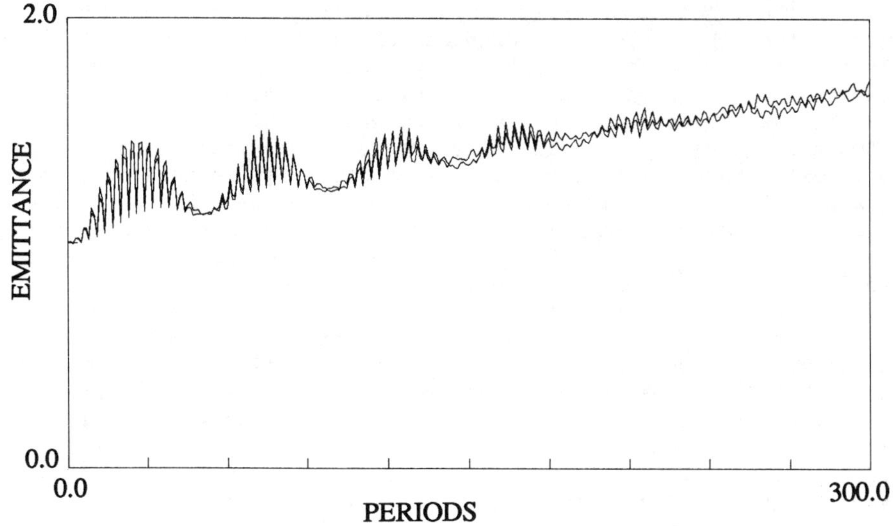

Fig. 7. Plot of the x and y rms emittances for the same parameters as in Fig. 6 but with a 256 x 256 grid, showing an almost halving of the emittance growth.

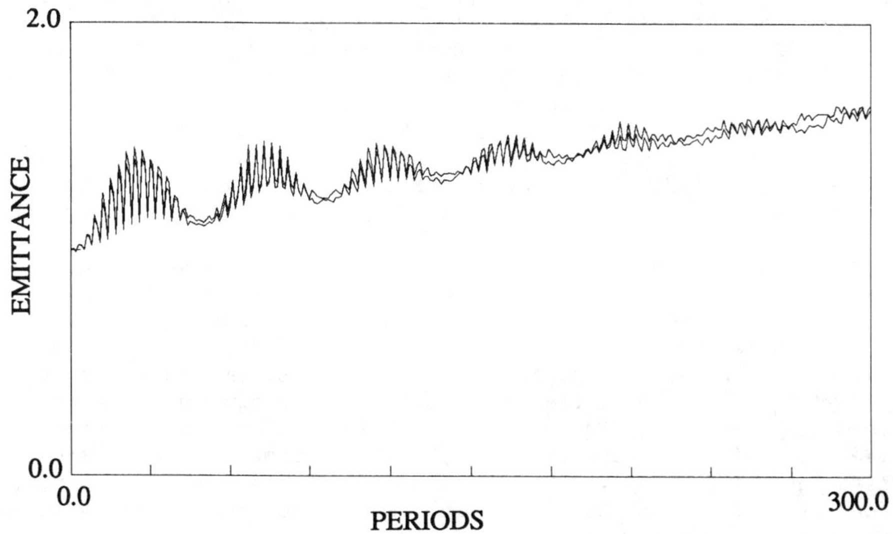

Fig. 8. Plot of the x and y rms emittances for a 512 x 512 grid showing almost no change from the 256 x 256 system in Fig. 7.

In an ensemble of fully ionized particles, the forces which result from the coulomb interactions among the particles will be "collisionless" when the number of particles per Debye sphere is larger than the number of plasma periods for which the system is to be followed. The number of particles whose orbits are followed in a simulation code is typically many orders of magnitude less than the physical plasma being modelled. Fluctuation fields arising from the granularity of the number density, which might be correct for the number of particles in the simulation, but not for the system being modelled, can therefore cause short scale distortions in phase space which will result in emittance growth.

This deviation from the collisionless behavior is often difficult to avoid using codes which sum the interactions between particles, because the n^2 scaling of the operation count limits the number of particles which can be practical to run. Since the collisionless behavior is a short wavelength phenomenon, it is possible to reduce collisions by smoothing the particle-particle interactions. This can be done in k-space in an FFT code, or by various grid averaging techniques on finite difference codes, and is automatically done by increasing the cutoff radius in many of the direct interaction codes. Such collision avoidance is obtained, however, at the cost of decreased resolution.

An example of the consequences of collisions has been obtained by running a thin lens AG system with phase advance of 90° per cell depressed to 30° by space charge. Figure 9, which was run with 16K particles, shows the evolution of an initial K-V distribution as the system goes unstable. The emittance remains constant during the linear phase of instability growth, then grows rapidly by about a factor of two and finally the growth saturates. Figure 10 is the same system, but run with 1024 particles. Some rounding of the break points in the curve can be seen but the major feature to note is the gradual upward slope which occurs after saturation. A further decrease in the number of particles further accentuates the rounding of the break points in the curve and the secular growth in the emittance, so that in the 128 particle simulation shown in Fig. 11, the structure of the region of rapid growth followed by saturation is almost gone. Since the collision frequency, and therefore the growth rate, is inversely proportional to the number of particles, Fig. 12 is a plot on a logarithmic scale of the secular emittance growth rate after saturation vs. number of particles. The straight line behavior expected is a fairly good representation of the data. However, the run with 16K particles lies somewhat above the straight line drawn through the points. The behavior of this emittance growth upon variation of other numerical and physical parameters, such as the beam intensity, have verified that this growth is due to a phenomenon that is probably not numerical. However the mechanism causing it is not yet well understood.

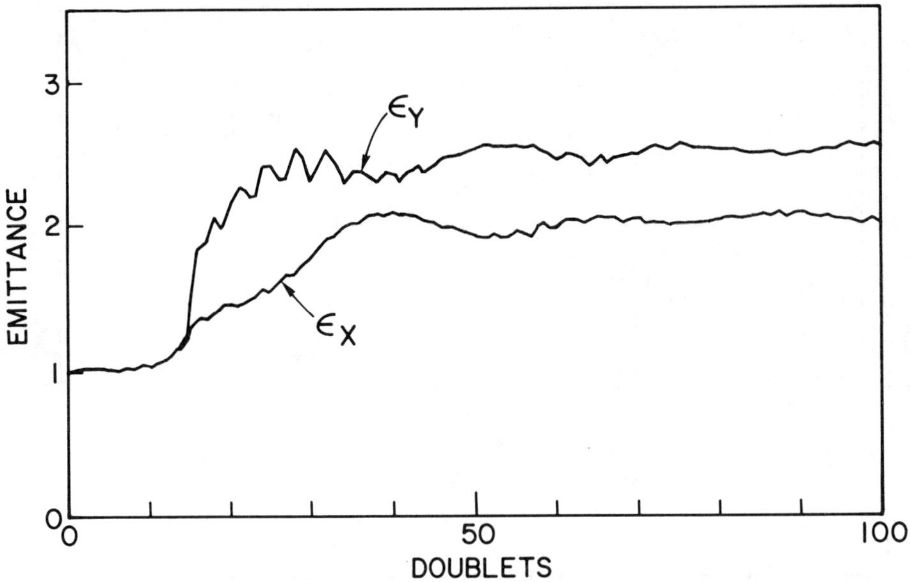

Fig. 9. Evolution of the x and y rms emittances of an initially K-V distribution with enough space charge to depress the 90° phase advance to 30°. This simulation employed 16 K particles.

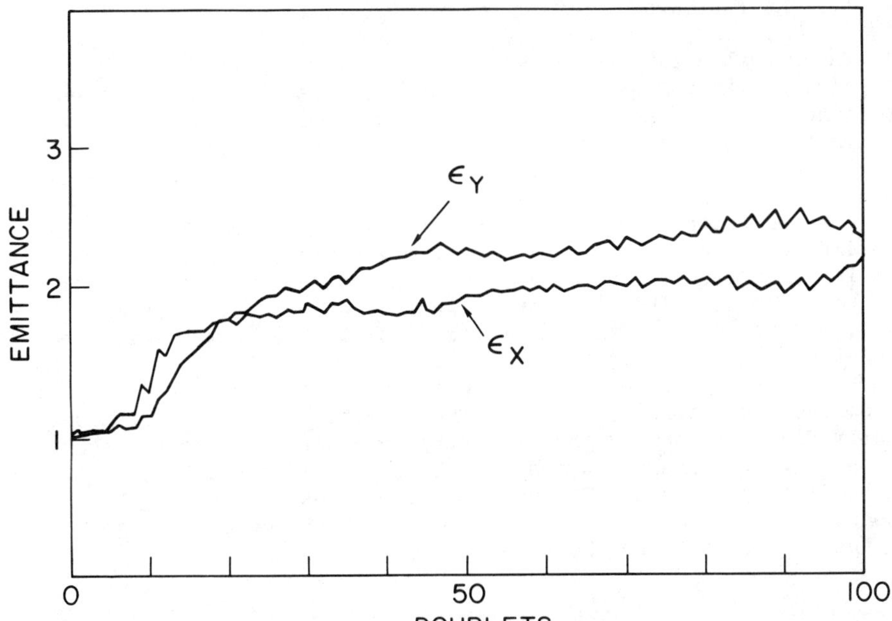

Fig. 10. Evolution of the emittances as in Fig. 9, but with 1024 particles, showing a rounding of the break points in the curve and some secular growth after saturation.

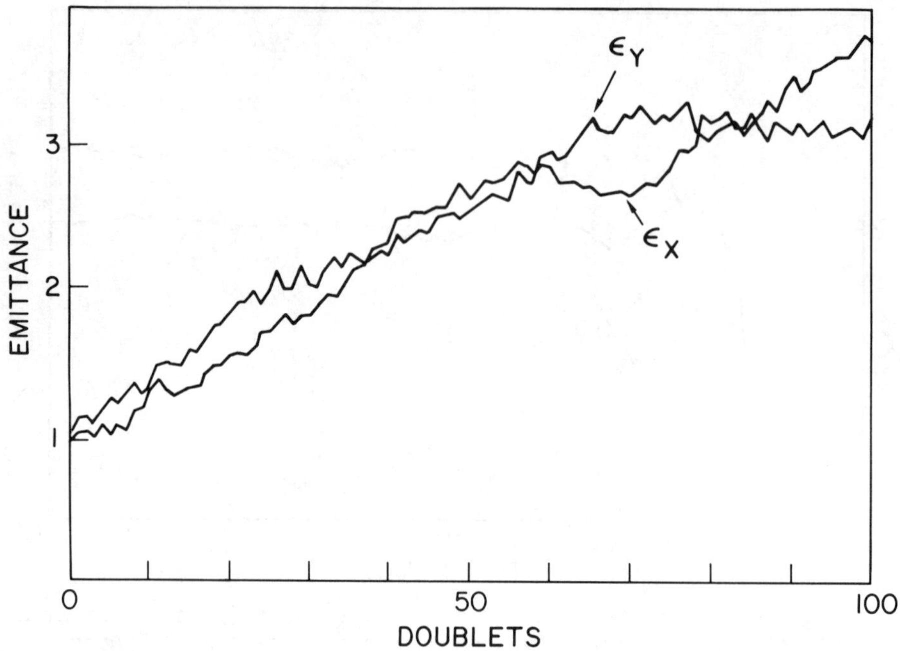

Fig. 11. Evolution of a system with 128 particles.

The actual relationship between collision frequency and emittance growth for any particular transport system has not been extensively investigated. A somewhat lower collisional growth rate has, in fact, been reported[27] for a different parameter regime and with a careful initial loading of the particles to minimize initial fluctuations.

The two examples presented thus far are characteristic of the behavior seen when moving with the beam and simulating the transverse behavior in the beam frame. The longitudinal dynamics of a beam, long compared with its diameter, however, can be quite different - especially if the beam thermal velocities, in the beam frame, are much slower than the plasma wave velocities in that frame. Such a cold beam is subject to a host of phenomena including wave steepening and soliton formation[12]. The timestep limits of such a system are usually fixed either by the necessity to resolve the particle and wave propagation transit times of a beam radius or the transverse limits discussed above. The spatial resolution in the longitudinal dimension is similarly fixed by the necessity to resolve longitudinal distances comparable to the beam radius.

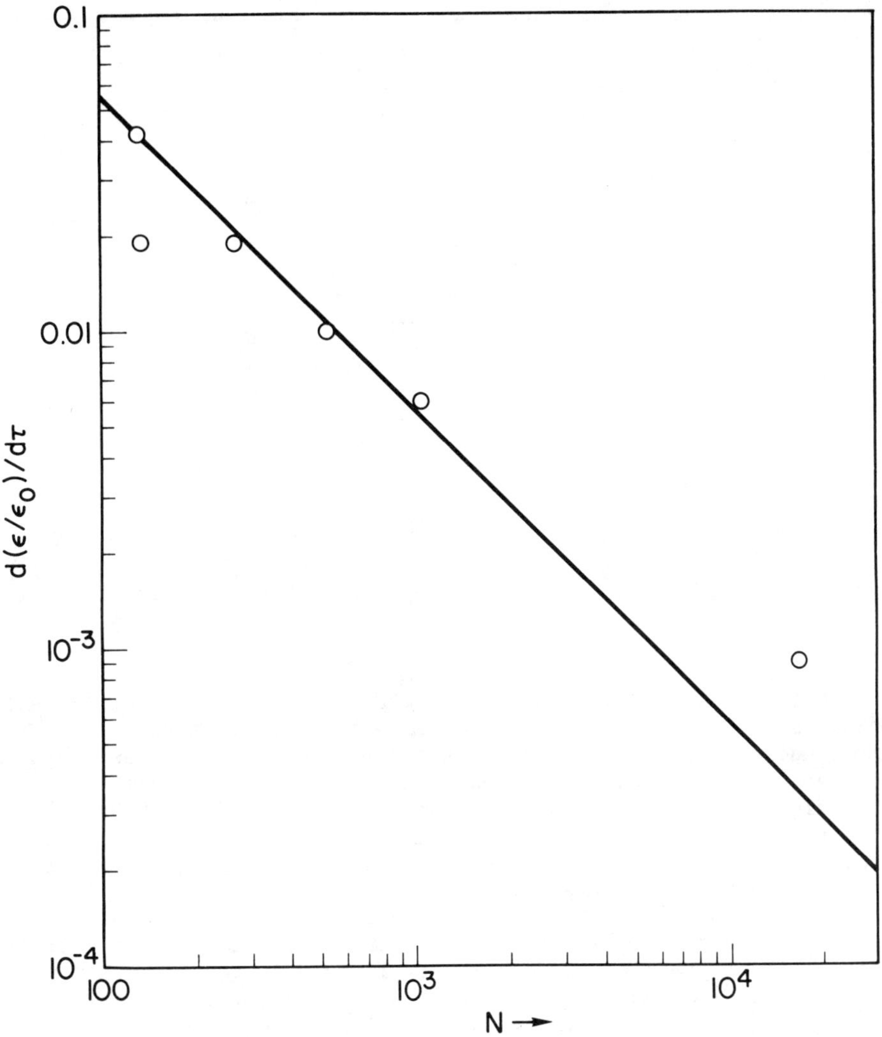

Fig. 12. Logarithmic plot of the emittance growth rate vs. number of particles in the simulation showing the approximate straight line dependence expected.

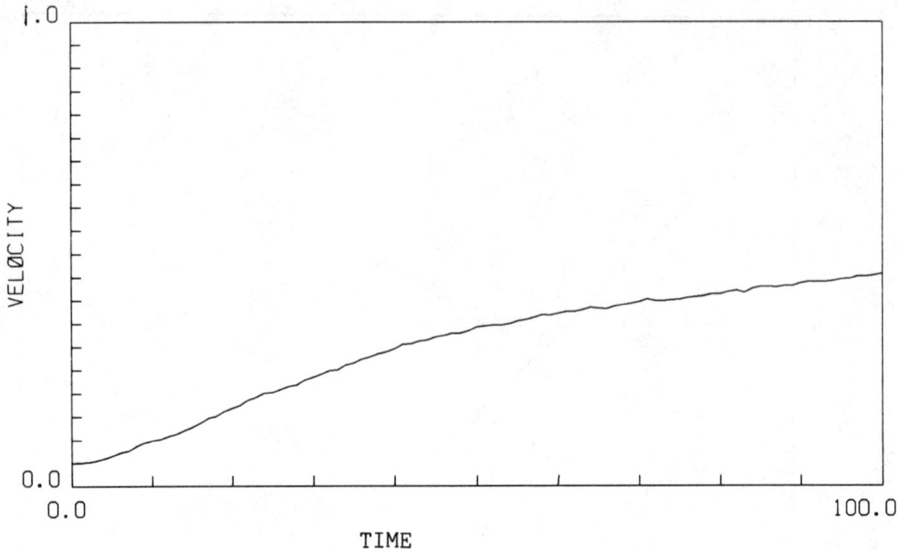

Fig. 13. Evolution of the rms velocity for a uniform longitudinal system with an initial thermal velocity of .001 times the plasma wave velocity.

Collisional effects in this longitudinal system can also present problems, particularly if it is desired to simulate a wave whose amplitude is greater than the beam thermal velocity but less than the wave velocity. Fluctuations resulting from the finite number of particles in such a system set up short wavelength fields which can trap particles and persist for some time. The heating and phase space structure which results can then mask any physics which is evolving on the same time scale.

Figure 13 is a plot of the evolution of the longitudinal rms velocity of a spatially uniform beam with an initial thermal velocity that is 0.001 times the wave velocity in that system. The numerical system used is one-dimensional, with an electric field that is proportional to the derivative of the line charge density, as is appropriate for a long beam in a conducting pipe. The system is 64 cells long and 2048 particles were used. The horizontal scale corresponds approximately to ten times the distance in cells that a plasma wave would propagate in that time.

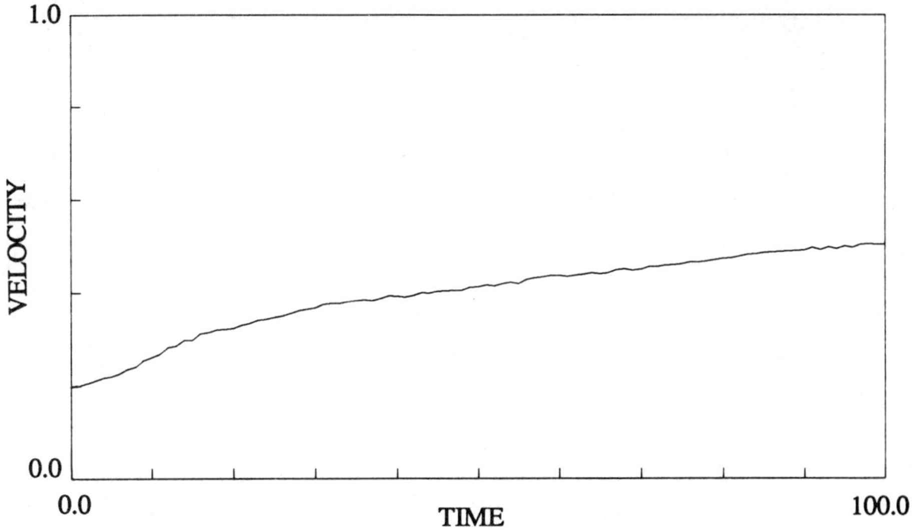

Fig. 14. Evolution of a system which is initially the same as in Fig. 13 but with five times the initial thermal velocity.

This behavior, though it is due to the finite number of particles in the simulation, is somewhat different from what is observed in the transverse simulations. The fine structure which develops in phase space and persists, seems to indicate that individual particles are being trapped in the fluctuation potential. This would suggest that, in contrast to the transverse behavior, either increasing the thermal velocity so that the fluctuation field amplitude is decreased, or increasing the thermal energy so that the particles are not trapped should almost completely alleviate the problem. Figure 15 shows the evolution of the thermal velocity when the initial thermal velocity is again increased by a factor of two, so that the starting thermal amplitude is comparable to the amplitude at the end of the first two runs. Emittance growth is still quite noticeable. However, a further increase of a factor of two in the initial thermal amplitude has a dramatic effect on the emittance growth. Figure 16 , shows the behavior for this system, which now has a starting amplitude above the saturation amplitude in the previous cases. Emittance growth has been reduced to about 1%. Note that the scale on this plot is twice that on the previous. Another factor of two increase in the thermal velocity reduces the growth somewhat further but the 1% growth may be acceptable for most purposes.

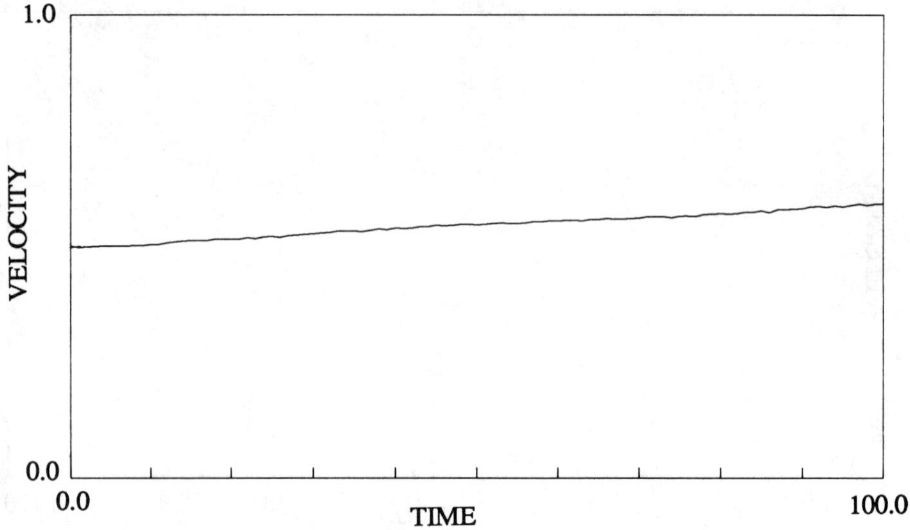

Fig. 15. Evolution of the rms velocity when the initial thermal velocity is increased another factor of two so that it is ten times the value in Fig. 13 and comparable to the final values in Figs. 13 and 14.

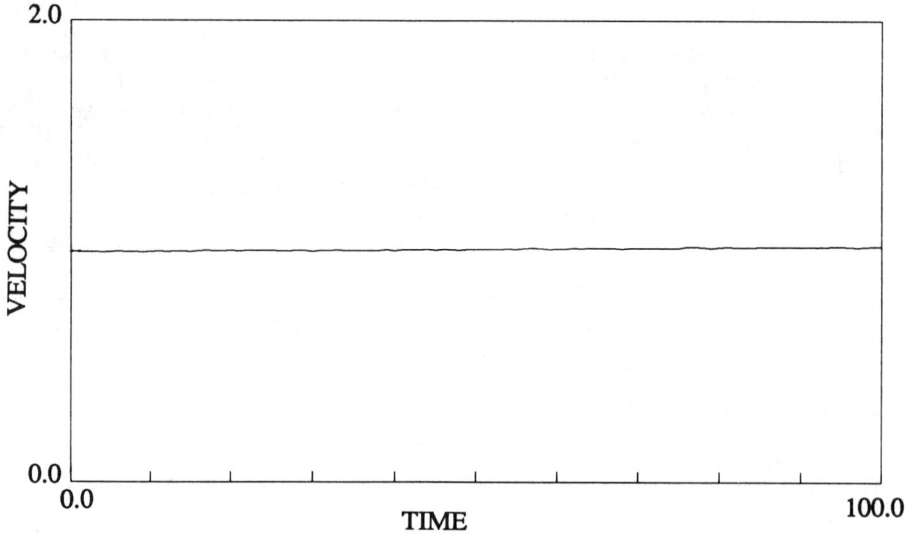

Fig. 16. Evolution of the rms velocity when the initial value is increased another factor of two. Note that the verticle scale on this plot is twice the previous.

If the thermal velocity is made very small, and the beam loaded in a regular manner, the onset of fluctuations can be delayed by the lack of any initial fluctuations until numerical errors in the simulation cause them to develop. Though not explicitly mentioned, all the simulations described here have been run with some k-space smoothing applied to the fields to attenuate the wavelengths of order of a numerical cell length. By multiplying the Fourier components of the fields by the factor $\exp(-k^2a^2)$, the system of equations behaves as if the individual particles are macroparticle clouds with a density distribution $\exp(-x^2/2a^2)$ in each dimension, where a is typically chosen to be one cell.

Figure 17 is a plot of the phase space of a longitudinal system with an initial wave perturbation and some instability but run with zero initial thermal velocity spread and no smoothing. A small initial long wavelength perturbation was introduced at the center of the system and it has grown as it propagates to the right, so that it is visible by this time. This perturbation should continue to grow until nonlinearities cause it to steepen and break. Figure 18 is a plot of the system a short time later. Since short wavelengths have slightly higher growth rates in this system, the initial perturbation has started to break up as the short wavelengths, introduced by numerical errors, overtake the amplitude of the long wavelength perturbation. Figure 19 shows this process continuing so that it is clear that the simulation results do not represent the physical

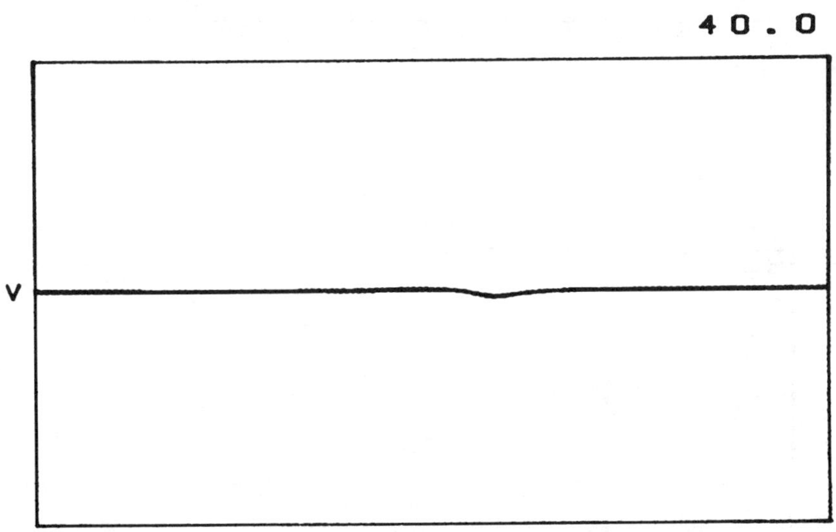

Fig. 17. Longitudinal phase space showing the propagation of a long wavelength perturbation in the presence of instability. System was initialized with no longitudinal temperature. No smoothing of short wavelength modes has been used.

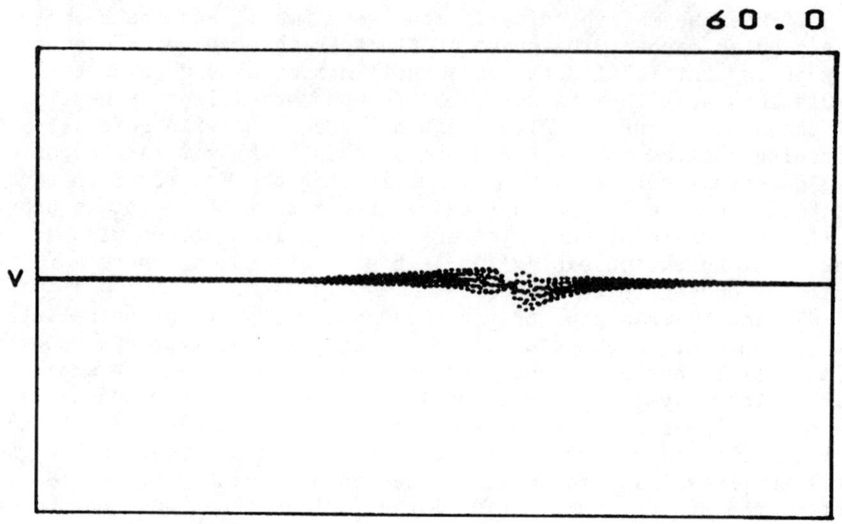

Fig. 18. Longitudinal phase space slightly later showing the rapid onset of anomalous short wavelength fluctuations.

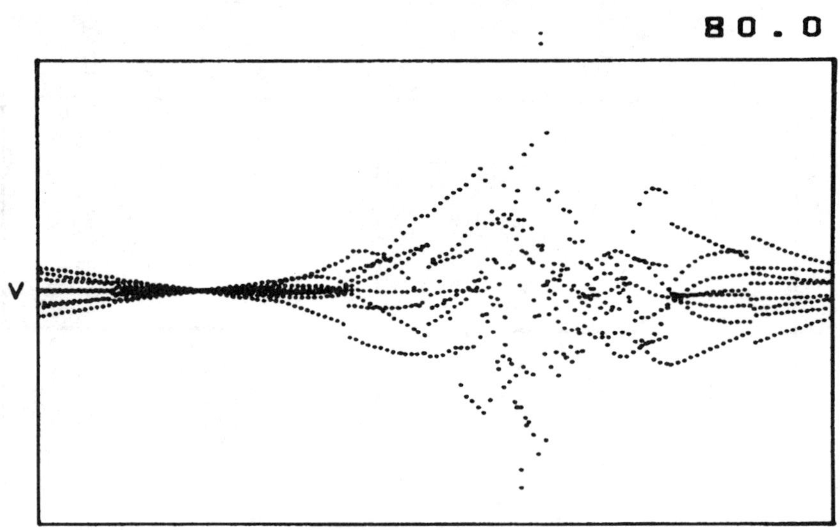

Fig. 19. Longitudinal phase space slightly later showing the continued rapid growth of anomalous fluctuations.

system which it is desired to simulate. The physical system, because it can have a very low thermal fluctuation level and the short wavelengths are damped, would in fact be expected to exhibit smooth growth of the long wavelength perturbation. This same physical system can, however, be run many tens of times as long without the appearance of the short wavelength modes if the smoothing is put back in. In that case the expected nonlinear growth and breaking of the initial perturbation is in fact observed.

CONCLUSIONS

Several examples have been cited of the use of particle in cell numerical algorithms to simulate physics which is relevant to the transport of intense ion beams. An attempt has been made to illustrate some instances in which the numerical parameters used in these simulations can affect the outcome. There are many instances, therefore, in which it is prudent to vary the numerical parameters of a simulation to insure that the results obtained are independent of these numerical parameters over a substantial range. In many instances this variation in numerical parameters is difficult because of machine time and memory limitations. It might then be prudent to examine a simplified model for which the numerical limitations are not as severe or to employ another numerical algorithm which, though not as suitable for running the particle physics to be simulated, can still nevertheless be used to benchmark the extent to which numerical limitations are affecting the desired outcome.

ACKNOWLEDGMENT

This work was supported by the United States Department of Energy under contracts DE-AI05-83ER40112 and DE-AI05-84ER40156.

REFERENCES

1. I. Hofmann, L.J. Laslett, L. Smith, I. Haber, "Stability of the Kapchinskij-Vladimirskij (K-V) Distribution in Long Periodic Transport Systems," Particle Accelerators, 13, 145 (1983).
2. A. W. Maschke, J. Brodowski and E. Meier, "Space Charge Limits in ESQ Transport Systems," IEEE Trans. on Nucl. Sci., NS-30, 2558 (1983).
3. M. G. Tiefenback and D. Keefe, "Measurements of Stability Limits for a Space-Charge-Dominated Ion Beam in a Long A.G. Transport Channel," Proceedings of the 1985 Particle Accelerator Conference, to be published in IEEE Trans. on Nucl. Sci., NS-32, (1985).
4. I. Haber, "Space Charge Limited Transport and Bunching of Non K-V Beams," IEEE Trans. on Nucl. Sci., NS-26, 3090 (1979).
5. I. Haber, "Simulation of Low Emittance Transport," Proceedings of the INS Symposium on Heavy Ion Accelerators and their Application to Inertial Fusion, Jan. 23-27, 1984, Institute for Nuclear Study, University of Tokyo, 451 (1984).

6. T. Wangler, "Field Energy and RMS Emittance," proceedings of this conference.
7. I. Hofmann, "Emittance Growth of Beams Close to the Space Charge Limit," IEEE Trans. on Nucl. Sci., NS-28, 2399 (1981).
8. C. M. Celata, I. Haber, L. J. Laslett, L. Smith, and M. G. Tiefenback, "The Effect of Induced Charge at Boundaries on Transverse Dynamics of a Space-Charge-Dominated Beam," Proceedings of the 1985 Particle Accelerator Conference, to be published in IEEE Trans. on Nucl. Sci., NS-32, (1985).
9. C. M. Celata, V. O. Brady, L. J. Laslett, L. Smith, "The Effect of Focusing Field Nonlinearities in MBE-4 on Transverse Beam Dynamics," op. cit.
10. H. Dantsker-Rudd, I. Haber, M. Reiser, "Simulation of Instabilities and Emittance Growth in a Periodic Solenoidal Channel with Linear and Nonlinear Focusing," op. cit.
11. C. R. Prior, "Multiturn Injection for Heavy Ion Fusion," Proceedings of the Symposium on Accelerator Aspects of Heavy Ion Fusion, GSI-82-8 Report, 290 (1982).
12. J. Bisognano, I. Haber, L. Smith, A. Sternlieb, "Non-Linear and Dispersive Effects in the Propagation and Growth of Longitudinal Waves on a Coasting Beam," IEEE Trans. on Sci., NS-28, 2513 (1981).
13. J. Bisognano, I. Haber, L. Smith, "Threshold Behavior for Longitudinal Stability of Induction Linac Bunches," IEEE Trans. on Nucl. Sci., NS-30, 2501 (1983).
14. I. Hofmann, I. Bozsik, and J. Jahnke, "Theoretical Study of High-Current Beams in Storage Rings," IEEE Trans. on Nucl. Sci. NS-30, 2546 (1983).
15. I. Hofmann, I. Boszik, "Multi-Dimensional Computer Simulations of Intense Beams," Proceedings of the Symposium on Accelerator Aspects of Heavy In Fusion, GSI-82-8 Report, 181 (1982).
16. I. Hofmann, I. Boszik, "Space Charge Problems in Final Beam Transport, Bunching and Focusing," op. cit. p. 362.
17. I. Haber, "Simulation of the Longitudinal Dynamics of a Cold Bunched Beam," op. cit. p.372.
18. J. Bisognano, E. P. Lee, J. W-K. Mark, "Numerical Studies of High Current Beam Compression in Heavy Ion Fusion," Proceedings of the 1985 Particle Accelerator Conf., to be published in IEEE Trans. on Nucl. Sci., NS-32, (1985).
19. Charles K. Birdsall and A. Bruce Langdon, Plasma Physics via Computer Simulation, (McGraw Hill, New York, 1985).
20. R. W. Hockney and J. W. Eastwood, Computer Simulation Using Particles, (McGraw Hill, New York, 1981).
21. J. P. Boris and R. A. Shanny, eds., Proceedings of the Fourth Conference on Numerical Simulation of Plasmas, Stock 0851 00059, (U.S. Govt. Printing Office, Washington DC, 1970).
22. B. Adler, S. Fernbach and M. Rotenberg, eds., Methods in Computational Physics, Controlled Fusion, 16, (Academic Press, New York, 1976).
23. B. Adler, S. Fernbach and M. Rotenberg, eds., Methods in Computational Physics, Plasma Physics, 9, (Academic Press, New York, 1970).

24. Ronald D. Ruth, "A Canonical Integration Technique," IEEE Trans. on Nucl. Sci., NS-30, 2669 (1983).
25. R. W. Hockney, "The Potential Calculation and Some Applications," in Methods of Computational Physics, ed. by B. Adler, S. Fernbach and M. Rotenberg (Academic Press, New York, 1970), 9, 162.
26. B. L. Buzbee, F.W. Dorr, J. A. George and G. H. Golub, "The Direct Solution of the Discrete Poisson Equation on Irregular Regions," SIAM J. Numr. Anal. 8, 722 (1971).
27. R. J. Hayden and M. J. Jakobsen, "Macrofilament Simulation of High Current Beam Transport," Proceedings of the 1985 Particle Accelerator Conference, to be published in IEEE Trans. on Nucl. Sci., NS-32 (1985).

FIELD ENERGY AND RMS EMITTANCE IN INTENSE PARTICLE BEAMS*

T. P. Wangler, K. R. Crandall, and R. S. Mills, MS H817
Los Alamos National Laboratory, Los Alamos, NM 87545

M. Reiser
University of Maryland, College Park, MD 20742

ABSTRACT

An equation is presented for continuous beams with azimuthal symmetry and continuous linear focusing; the equation expresses a relationship between the rate of change for squared rms emittance and the rate of change for a quantity we call the nonlinear field energy. The nonlinear field energy depends on the shape of the charge distribution and corresponds to the residual field energy possessed by beams with nonuniform charge distributions. The equation can be integrated for the case of an rms matched beam to yield a formula for space-charge-induced emittance growth that we have tested numerically for a variety of initial distributions. The results provide a framework for discussing the scaling of rms emittance growth and an explanation for the well-established lower limit on output emittance.

INTRODUCTION

A quantitative understanding of emittance growth induced by space charge is of great importance for the design of high-current linear accelerators and beam transport systems. The first systematic numerical study of emittance growth for linac beams was done by Chasman,[1] and more recent studies were reported by Jameson and Mills.[2] An interest in beam instabilities of periodic transport channels was stimulated by heavy-ion fusion requirements, and led to work recently reported by Hofmann, Laslett, Smith, and Haber.[3] The mechanism of equipartitioning and emittance growth in linacs was studied by Hofmann and Bozsik[4] and also by Jameson.[5] Reviews of this work have been presented by Jameson[6] and Hofmann.[7]

In spite of these efforts, basic phenomena observed both in numerical studies and in unpublished experimental data have remained unexplained, and no clear explanation of many of the important effects has emerged. In the past little attention was given to a comparison of rms emittance growth for different initial distributions. Recently, numerical studies were published by Struckmeier, Klabunde, and Reiser[8] for a continuous beam in a periodic channel, which led them to observe that some distributions experience a rapid initial emittance growth that increases with beam intensity. These authors hypothesized that the observed emittance growth was associated with a homogenization of the charge

*Work supported by the US Department of Energy.

density and resulted in a conversion of space-charge field energy into transverse particle energy. They suggested that transverse energy conservation could be used to obtain a useful formula for emittance growth. Two formulas were presented, which were interpreted as upper and lower bounds on the emittance growth.

The idea of space-charge field energy as a useful quantity has been suggested before. In 1970, Gluckstern, et al.[9] used electric-field energy comparisons to investigate the relative stability of different stationary distributions. Electric-field energy relationships were considered in more detail by Lapostolle[10,11] in his important work in 1971, which, together with work by Sacherer,[12] resulted in a generalized rms envelope equation. Thus, stimulated by the suggestions of Struckmeier, Klabunde, and Reiser, we have attempted to extend the initial work of Lapostolle and Sacherer and to investigate further the relationship between space-charge field energy and rms emittance for continuous beams in continuous focusing channels.

RMS EMITTANCE AND NONLINEAR FIELD ENERGY

We consider the problem of a continuous, azimuthally symmetric beam that propagates in the +z-direction at constant velocity v. We assume that the beam is confined radially by a continuous, linear, external focusing force and that the paraxial approximation is valid in which all particles have the same longitudinal velocity $v \gg v_t$, where v_t is the transverse velocity component. We consider the characteristics of a steady-state solution, where the charge density, current density, and fields have no explicit dependence on time. We allow for initial phase-space distributions that are not necessarily stationary, and in general, we expect that the charge density $\rho(r,z)$ will evolve from an initial state at $z = 0$ to some final state, which may be stationary or independent of z.

Maxwell's equations can be written using nonzero self-field components E_r, E_z, and B_θ, and current-density components j_r and j_z. A solution is easily obtained in the approximation that $\partial E_z/\partial z \ll \rho/\varepsilon_0$, where ε_0 is the free-space permittivity, which leads to

$$B_\theta(r,z) = \frac{v}{c^2} E_r(r,z) \quad , \tag{1}$$

and

$$j_r(r,z) = -\varepsilon_0 v \frac{\partial E_r(r,z)}{\partial z} \quad . \tag{2}$$

The component $E_z(r,z)$ is not ignored, but is determined by the requirement that $\oint \vec{E}\cdot d\vec{s} = 0$, and by a boundary condition such as a perfectly conducting pipe that encloses the beam.

The second moments of the charge distribution in x,x' phase space are $\overline{x^2}$, $\overline{x'^2}$, and $\overline{xx'}$, where the bar indicates an average over the phase-space distribution. It can be shown that the derivatives of the moments are functions of the moments and of the values of $\overline{xF_x}$ and $\overline{x'F_x}$, where F_x is the x-component of the total force, the sum of the external plus the self-force.[11,12] The definition of rms emittance ε in x,x' phase-space, given by Lapostolle, can be shown to correspond to the total emittance of a continuous equivalent uniform beam (same second moments as the real beam) and is given by

$$\varepsilon = 4(\overline{x^2}\,\overline{x'^2} - \overline{xx'}^2)^{1/2} \quad . \tag{3}$$

The rms emittance represents a convenient measure of the macroscopic or effective emittance of a beam subjected to nonlinear forces that arise either from external or self-fields.[13] If Eq. (3) is differentiated and if the external force is assumed to be linear, it can be shown that

$$\frac{d\varepsilon^2}{dz} = \frac{32}{m\gamma v^2}(\overline{x^2}\,\overline{x'F_{sx}} - \overline{xx'}\,\overline{xF_{sx}}) \quad , \tag{4}$$

where m is the mass, and $\gamma = (1 - v^2/c^2)^{-1}$. For beam particles with charge e, $F_{sx} = eE_x/\gamma^2$ is the x-component of the self-force, including both electric and magnetic terms, which have opposite signs. Sacherer derived $\overline{xF_{sx}}$ for an arbitrary charge distribution with elliptical symmetry, and his result can be written

$$\overline{xF_s} = 2w_o/N\gamma^2 \quad , \tag{5}$$

where N is number of particles per unit length and

$$w_o = (eN)^2/16\pi\varepsilon_o \quad . \tag{6}$$

The value of $\overline{x'F_{sx}}$ can be derived using Eq. (2), which leads to

$$\overline{x'F_{sx}} = -\frac{1}{2N\gamma^2}\frac{dW}{dz} \quad , \tag{7}$$

where

$$W = \pi\varepsilon_o \int_0^\infty r\,E_r^2\,dr \tag{8}$$

is the self-electric field energy per unit length associated with the radial electric field. With an infinite value for the upper limit to the integral of Eq. (8), we find that the field energy W diverges. However, only field-energy changes or field-energy differences will be important. Therefore, because the electric field outside the beam is independent of the charge density, the infinite upper limit can be replaced by any finite radius larger than the actual beam. For a uniform beam, we write the field energy as W_u, and for a volume within a radius b (larger than the beam) we obtain

$$W_u = w_o(1 + 4 \ln b/X) \quad , \quad \text{and} \quad b \geq X \quad , \tag{9}$$

where X, defined by $X = 2\sqrt{\overline{x^2}}$ is the total beam radius for a uniform beam. Using these results, Eq. (4) can be written as

$$\frac{d\varepsilon^2}{dz} = -\frac{X^2 K}{2}\frac{d}{dz}\left(\frac{U}{w_o}\right) \quad , \tag{10}$$

where X (twice the rms beam size) is interpreted as the total beam radius of the equivalent uniform beam, K is the generalized perveance given by

$$K = \frac{eI}{2\pi\varepsilon_o m v^3 \gamma^3} \quad , \tag{11}$$

and

$$U = W - W_u \quad . \tag{12}$$

We learned recently that a form equivalent to Eq. (10) was included by Lapostolle in Refs. 10 and 11. Instead of U/w_o, Lapostolle's equation uses a field-energy correction term ε_2, which can be shown to be linearly related to U/w_o. We also learned that a similar equation was obtained by Lee, Yu, and Barletta[14] for a beam with no external focusing.

The quantity U is the difference between the self-electric field energies per unit length of the distribution and of the equivalent uniform beam. The special role of the uniform distribution can be explained by its associated linear self-force, which causes no rms emittance growth. Thus, U is the residual self-electric field energy possessed by beams with nonuniform charge distributions. Because nonuniform beams have nonlinear self-fields, we call U the nonlinear field energy. Equation (10) implies that a decrease in nonlinear field energy U corresponds to an increase in rms emittance. Both the electric- and magnetic-field contributions are contained in Eq. (10) by including the factor γ^3 in the definition of K (γ^2 accounts for the magnetic field and γ accounts for relativistic mass). Although the total electromagnetic stored energy is the sum of electric plus magnetic terms, it is the difference that is related to changes in transverse rms emittance because the transverse electric and magnetic self-force terms have opposite signs. Therefore, rms emittance growth, induced by space charge, is inherently a nonrelativistic effect; it is most important when γ is near unity.

From Eq. (9), w_0 is the self-electric field energy per unit length within the beam boundary of an equivalent uniform beam. The quantity U/w_0 is dimensionless, and its properties can easily be demonstrated by considering the example of a power-law charge distribution. We find that U/w_0 is zero for a uniform charge distribution and is positive both for peaked and hollow distributions, increasing as the distribution becomes more nonuniform. Furthermore, U/w_0 is independent of both beam current and rms beam size and is a function only of the shape of the distribution. Thus, Eq. (10) shows that rms emittance changes are associated with three separate factors: rms beam size, perveance, and changes in shape of the charge distribution. Values of U/w_0 are given for distributions as shown in Table I. These distributions are listed in order from most peaked to most hollow.

TABLE I

U/w_0 FOR SOME COMMON DISTRIBUTIONS

Spatial Distribution	Charge Density $\rho(r)$		U/w_0
Gaussian	$\exp(-r^2/a)$		0.154
Waterbag	$1 - (r/R)^2$	$r \leq R$	0.0224
Uniform	1	$r \leq R$	0.000
Hollow (n = 2)	r^2	$r \leq R$	0.0754
Hollow (n = 10)	r^{10}	$r \leq R$	0.245

Equation (10) can be easily integrated, if we assume an rms-matched beam with constant X. The resulting relation between rms emittance and nonlinear field energy can be expressed as

$$\frac{\varepsilon}{\varepsilon_i} = \left[1 - \frac{(U - U_i)}{2w_0}\left(\frac{\omega_0^2}{\omega_i^2} - 1\right)\right]^{1/2} , \qquad (13)$$

where ω_0 is the zero-current betatron frequency, ω_i is the initial betatron frequency for the equivalent uniform beam (including space charge), and ε_i and U_i are initial values of rms emittance and nonlinear field energy. In Eq. (13), the rms emittance is expressed as the product of two factors: one, $(U - U_i)/w_0$, related to the change in the shape of the distribution and one related to the initial betatron tune ratio ω_i/ω_0. It is easy to show that the tune ratio is $\omega_i/\omega_0 = \sqrt{1 + u^2} - u$, where the space-charge parameter $u = Kv/2\varepsilon_i\omega_0$.

We also find that an expression for transverse energy conservation can be derived using the assumptions given above. We obtain $\overline{T} + \overline{V_e} + W/N\gamma^2$ = constant, where $\overline{V_e}$ is the average potential energy associated with the external force, and the average transverse kinetic energy is $\overline{T} = m\gamma v^2\overline{x'^2}$ for an azimuthally symmetric beam in the paraxial approximation. In the limit of a perfect rms matched beam, $\overline{V_e}$ and W_u are constant, so that a decrease in U (and therefore W) corresponds to an increase in $\overline{T}$. In general, an initial rms-matched beam will not remain exactly matched if the rms emittance grows.

NUMERICAL STUDIES

We have made numerical studies using 4000 beam particles per calculation to investigate the effects described above, using a computer code written especially for this problem. We have studied azimuthally symmetric beams with different rms-matched initial distributions. At each time step, the particles were given transverse deflections based on both the external- and self-forces. The radial self-forces were calculated from Gauss' law, assuming infinitely long cylindrical charge distributions. The objectives were to study the evolution of the charge distribution, U/w_0, and ε for different initial distributions and different tune ratios ω_i/ω_0.

First, we present results for a space-charge-dominated beam with initial Gaussian distributions, both in position and divergence, truncated at two standard deviations and with an initial tune ratio $\omega_i/\omega_0 = 0.02$. We have introduced a radial distribution parameter, defined as $\rho = 1 - \overline{r}/\overline{r_u}$, where $\overline{r}$ is the average radius and $\overline{r_u}$ is the average radius for the equivalent (same second moments) uniform beam. For simple distributions, we find that ρ is positive for a peaked charge distribution, zero for a uniform beam, and negative for a hollow beam. However, the parameter ρ must be interpreted with caution because in some cases the distributions are too complicated for such a simple characterization. Figures 1a and 1b show ρ and U/w_0 as a function of distance z/λ_p along the beamline, where a plasma length, $\lambda_p = (2\pi^2X^2/K\gamma^2)^{1/2}$, is the distance the beam travels during one plasma period. The parameter ρ indicates that the beam distribution undergoes damped radial oscillations between peaked and hollow configurations at the plasma frequency, a result that is confirmed by a more detailed examination of the charge distributions.

The quantity U/w_0 is maximum at both the extreme peaked and hollow configurations and therefore oscillates at twice the frequency of the parameter ρ. The minimum value of U/w_0 during these oscillations generally is not the zero value expected for an exactly uniform charge distribution. We believe that two effects contribute to the nonzero minimum values of U/w_0: a low-density halo that remains outside the central core of the beam, and a numerical error caused by statistical fluctuations in the charge distribution. Statistical fluctuations also appear to contribute to some of the details of the curves in Fig. 1, beyond about five plasma oscillations. Figure 1c shows the emittance ratio $\varepsilon/\varepsilon_i$ as a function of z/λ_p. Two overlapping curves are shown: one obtained from Eq. (3) with an evaluation of second moments and the other from evaluation of U and using Eq. (13). The two curves are in excellent agreement and show a rapid initial

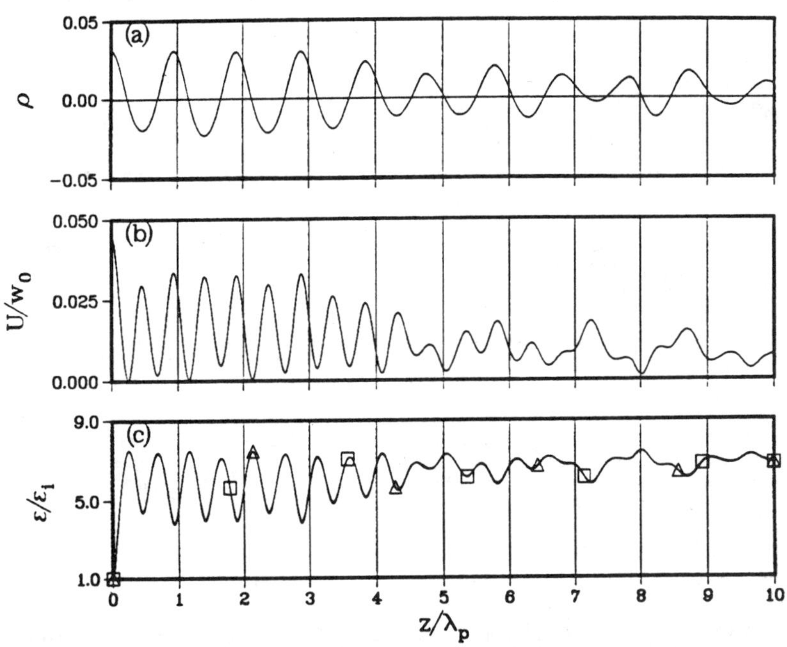

Fig. 1. Results of numerical simulation for 10 plasma periods using an initial Gaussian charge distribution, truncated at two standard deviations, and initial tune ratio $\omega_i/\omega_0 = 0.02$. The abscissa is the distance z/λ_p along the beamline, where λ_p is the distance the beam travels in one plasma period.
(a) Dimensionless radial distribution parameter ρ, defined in the text, is positive for a peaked charge distribution, zero for a uniform beam, and negative for a hollow beam.
(b) Dimensionless nonlinear field energy U/w_0.
(c) Emittance ratio $\varepsilon/\varepsilon_i$. There are two overlapping curves. The square symbols correspond to emittance calculated from second moments at each step, using Eq. (3). The triangles correspond to emittance calculated from nonlinear field energy at each step, using Eq. (13).

growth that occurs during the first quarter of the first plasma period. A closer examination shows that the two curves are not in exact agreement because of a slight mismatch caused by the emittance growth, an effect that is not included in the derivation of Eq. (13). Results for the same quantities, shown in Fig. 2 for 100 plasma periods, suggest that after about 10 to 15 periods the beam has evolved to an approximately stationary state. We believe that much of the residual noise seen in Fig. 2 is caused by statistical fluctuations.

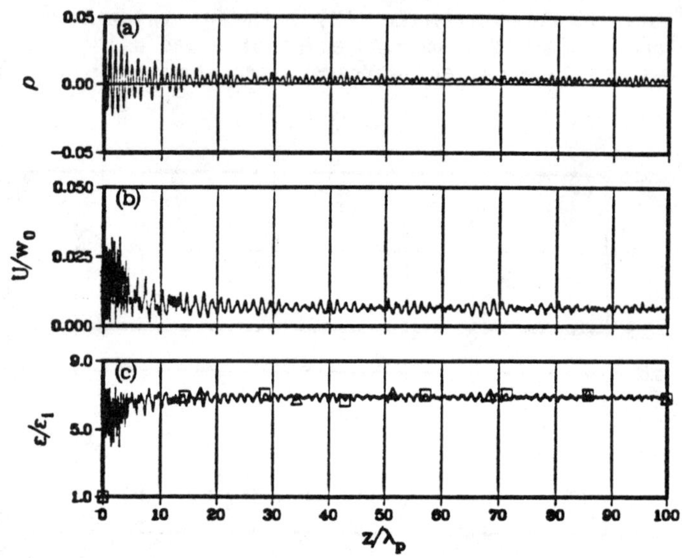

Fig. 2. Results of numerical simulation for 100 plasma periods using an initial Gaussian charge distribution, truncated at two standard deviations, and initial tune ratio $\omega_i/\omega_0 = 0.02$. The quantities plotted are the same as in Fig. 1.

Figure 3 shows x,x' phase-space, the x,y cross section, and the radial charge density at $z/\lambda_p = 0.0$, 0.25, 0.5, and 100. The variation of the phase-space distribution, seen in Figs. 3a through 3c ($z/\lambda_p = 0.0$, 0.25 and 0.5), appears to be qualitatively consistent with the rms emittance curve shown in Fig. 1c. The phase-space distribution at $z/\lambda_p = 100$, shown in Fig. 3d, appears to have evolved to a central core and an extended diffuse background and corresponds to an rms emittance growth ratio of $\varepsilon/\varepsilon_i = 6.7$. Figures 3e through 3g and 3i through 3k show that the charge density changes from an initial truncated Gaussian distribution at $z/\gamma_p = 0.0$ to a uniform distribution at $z/\lambda_p = 0.25$ and to a hollow beam at $z/\lambda_p = 0.50$. The charge density at $z/\lambda_p = 100$ can be seen in Figs. 3h and 3l and within the statistical uncertainties is consistent with a uniform distribution with a soft edge or finite thickness boundary. Evidence can be seen for a low-density halo outside the central beam.

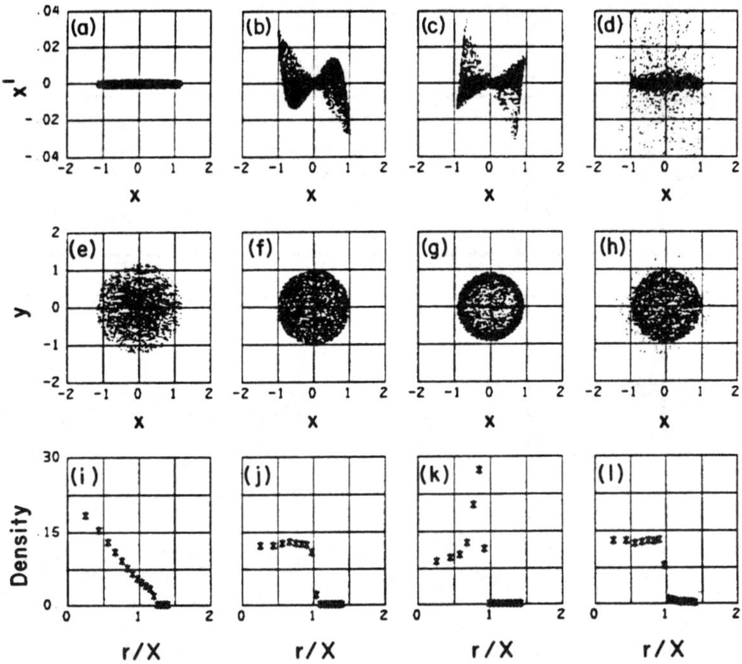

Fig. 3. Phase-space (x,x'), cross section (x,y), and charge density versus normalized radius r/X at z/λ_p = 0.0, 0.25, 0.50, and 100 from numerical simulation for an initial Gaussian distribution, truncated at two standard deviations, and initial tune ratio ω_i/ω_0 = 0.02. The quantities x, y, x' and charge density are shown in relative units. More discussion of these plots is given in the text.

A second numerical example is shown in Fig. 4 for a beam with an initial semi-Gaussian or thermal distribution, corresponding to uniform charge density, Gaussian distribution in velocity space (truncated at four standard deviations), and with an initial tune ratio ω_i/ω_0 = 0.25. Figures 4a and 4b show ρ and U/w_0 versus z/λ_p. The parameter ρ rises from its initial value of zero to remain positive throughout, indicating a peaked distribution. An examination of the charge distribution reveals that the initial hard-edged beam evolved rapidly to one with a tail or soft edge. The quantity U/w_0 increases, which would imply an emittance decrease according to Eq. (13). Figure 4c shows the rms emittance ratio $\varepsilon/\varepsilon_i$. Again, the two overlapping curves, one from Eq. (3) and the other from Eq. (13), are in excellent agreement, and both show a small decrease in the rms emittance.

Thus, we conclude that not only can the rms emittance increase as the beam becomes more uniform, but also an rms emittance decrease is possible when the beam becomes less uniform. As in the previous example, damped oscillations are observed, but at present, we have no quantitative explanation of oscillation frequencies for beams that are not highly space-charge dominated.

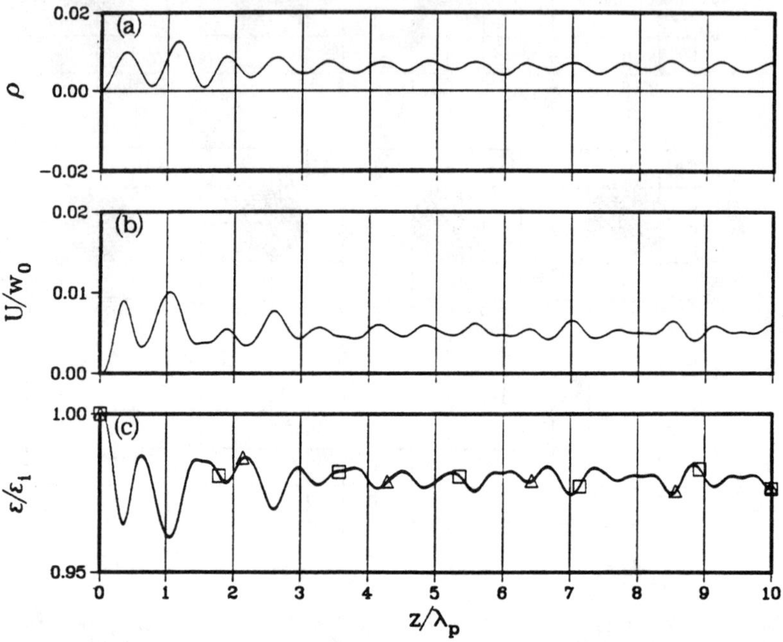

Fig. 4. Results of numerical simulation for an initial semi-Gaussian or thermal distribution, truncated at four standard deviations in velocity space, and initial tune ratio ω_i/ω_0 = 0.25. The quantities plotted are the same as in Fig. 1.

Numerical studies have been made for other initial distributions and tune ratios and for distances up to 100 plasma lengths. Our interpretation of these results is that the initially rms-matched beam undergoes damped radial oscillations and evolves to a final state with a central core and sometimes with a low-density halo that contains a few per cent of the beam. For all initial charge distributions, the final charge distribution is observed qualitatively to change from a Gaussian-like profile to one that is uniform with a hard edge, as the tune ratio decreases from ω_i/ω_0 = 1 to 0. This behavior can be interpreted as a decrease in thickness of the beam boundary and is reasonable physically because the thickness should be of the order of the Debye length, which increases monotonically with tune ratio. The final charge density and final U/w_0 depend on the initial phase-space distribution and on the initial tune ratio. Therefore, Eq. (13) implies that the rms emittance growth ratio for a given initial phase-space distribution is a function only of the initial tune ratio.

Equations (10) and (13) describe rms emittance variation caused by any change in U/w_0 (associated with a change in the shape of the charge distribution) regardless of the detailed mechanism. In particular, unstable oscillation modes of the charge density, which are a well-known property of the Kapchinskii-Vladimirskii (K-V) distribution below certain tune-ratio thresholds,[15] can cause such changes in charge density. The effects of the K-V instabilities leading to no growth of rms emittance have been reported by Hofmann.[16] It is interesting to reconsider this problem from the viewpoint of rms emittance and nonlinear field energy.

For an initial K-V uniform beam ($U_i/w_0 = 0$), any final charge distribution that is not uniform will result in an increase in U/w_0 and, according to Eq. (13), a decrease in rms emittance. Figure 5 shows U/w_0 after 40 plasma periods versus initial tune ratio from numerical simulations for an initial K-V distribution, using 4000 particles. Although the points are affected by statistical uncertainties, Fig. 5 shows that the final U/w_0 remains near zero for tune ratios $\omega_i/\omega_0 \gtrsim 0.4$. For lower tune ratios, we observe a peak in U/w_0, centered at about $\omega_i/\omega_0 = 0.3$. These results are consistent with the theory of instabilities for the K-V distribution,[15] which predicts that the beam is stable for $\omega_i/\omega_0 > 0.4$ and that many coherent oscillation modes have instability thresholds at tune ratios below $\omega_i/\omega_0 = 0.4$. The small final U/w_0 values observed for tune ratios near $\omega_i/\omega_0 = 0$ are consistent with a final uniform charge density that is expected in the extreme space-charge limit to obtain complete shielding of the linear applied external force (zero Debye length).

Figure 6a shows the initial charge distribution and Figs. 6b through 6d show the charge distribution after 40 plasma periods for $\omega_i/\omega_0 = 0.02$, 0.30, and 0.70, respectively. An examination of Fig. 6c suggests that the increase of U/w_0 near $\omega_i/\omega_0 = 0.3$ is caused by an increase in the thickness of the beam boundary. The existence of the peak in U/w_0 can be interpreted as the result of an increase in the final U/w_0 with respect to ω_i/ω_0, for the final stationary states of an unstable K-V beam.

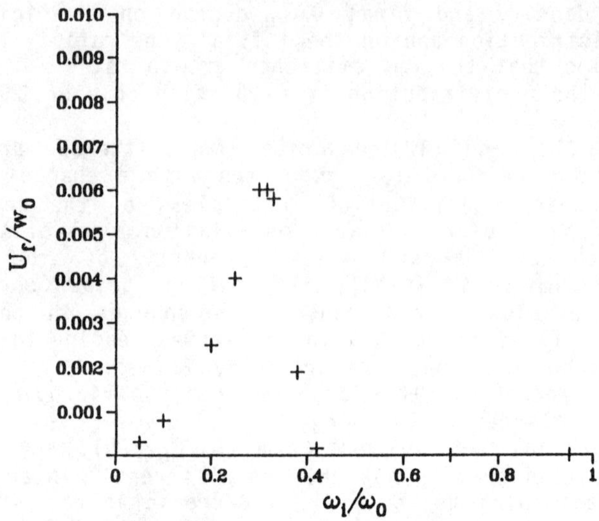

Fig. 5. Final U/w_0 versus initial tune ratio ω_i/ω_0 from numerical simulations for an initial K-V distribution.

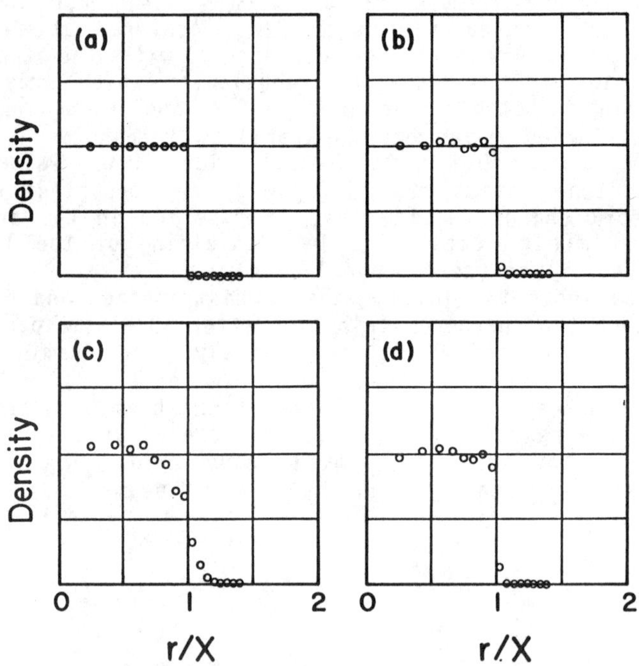

Fig. 6. Charge density in relative units versus normalized radius r/X after 40 plasma periods from numerical simulations for an initial K-V distribution. Figure 6a shows the initial charge density, and Figs. 6b through 6d show the final charge density at initial tune ratios ω_i/ω_0 = 0.02, 0.3, and 0.70.

Figure 7 shows the final emittance growth ratio versus initial tune ratio from the same particle simulations. For tune ratios $\omega_i/\omega_0 \lesssim 0.4$ a slight emittance decrease is observed, consistent with the predictions of Eq. (13). Thus, the absence of rms emittance growth for an unstable K-V beam in a continuous focusing channel can be explained as an absence of available space-charge field energy required for such growth.

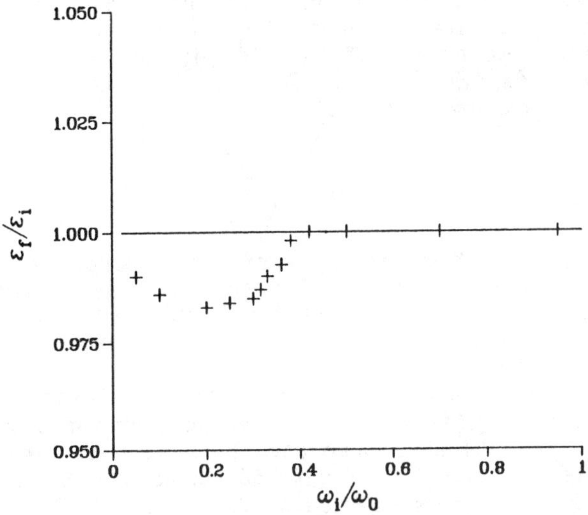

Fig. 7. Final emittance ratio $\varepsilon_f/\varepsilon_i$ versus initial tune ratio ω_i/ω_0 from numerical simulations for an initial K-V distribution.

FINAL, UNIFORM CHARGE-DENSITY APPROXIMATION

In spite of the excellent agreement of Eq. (13) with the numerical simulation results, we are unable to predict the final rms emittance growth unless the final value of U/w_0 is known. In the extreme space-charge limit when ω_i and ε_i approach zero, we expect that the beam will evolve towards a final stationary state with a uniform charge distribution to obtain complete shielding of the linear applied focusing force. As an approximation for all initial tune ratios, we will assume that in the final state, $U/w_0 = 0$, which corresponds to a uniform beam with no halo. This approximation should be good in the space-charge-dominated limit, where the emittance growth is largest; in the emittance-dominated limit, where the approximation could lead to a large error in emittance growth, the emittance growth itself is small. In all cases, the approximation will result in an overestimation of the emittance growth.

With this approximation, the final emittance for an rms-matched beam can be written directly from Eq. (13) as

$$\frac{\varepsilon_f}{\varepsilon_i} \simeq \left[1 + \frac{U_i}{2w_o}\left(\frac{\omega_o^2}{\omega_i^2} - 1\right)\right]^{1/2} . \tag{14}$$

This equation corresponds to the upper limit emittance growth formula, derived by Struckmeier, Klabunde, and Reiser,[8] if betatron frequencies are replaced by phase advances per quadrupole focusing period. We note that Eq. (14) depends on the properties of the initial beam, which can be easily calculated.

For a space-charge-dominated beam, Eq. (14) can be written as

$$\varepsilon_f^2 \cong \varepsilon_i^2 + \frac{1}{2}\left(\frac{Kv}{\omega_o}\right)^2 \left(\frac{U_i}{w_o}\right) . \tag{15}$$

This result implies that as ε_i approaches zero, the final emittance approaches a minimum value that decreases with increased focusing force (larger ω_0), increases with perveance K, and increases with initial nonuniformity as measured by U_i/w_0. This formula provides a quantitative description of the familiar lower limit on final emittance similar to that first reported in numerical studies by Chasman[1] for linear accelerator beams.

Numerical results have been used to test Eqs. (14) and (15). As an example, Figs. 8a through 8d show $\varepsilon_f/\varepsilon_i$ versus ω_i/ω_0 plotted on the same scale for initial Gaussian (truncated at four standard deviations), waterbag, thermal or semi-Gaussian (uniform charge density), and hollow (n = 2) distributions, whose initial U/w_0 values are given in Table 1. Values from the particle simulations are plotted and are in good agreement with the curves from Eq. (14). Therefore, we conclude that final emittance growth can be approximately predicted in advance from a knowledge only of the initial tune ratio and the initial value of U/w_0. A comparison of Eq. (15) with particle simulations is shown in Figs. 9a through 9d, where ε_f versus ε_i is plotted on the same scale for the same four distributions at a value of $Kv/\omega_0 = 5.0 \times 10^{-5}$ m. The results from the simulations and the curve from Eq. (15) are, again, in close agreement. The dependence of these results on the initial charge distribution is evident in Figs. 8 and 9.

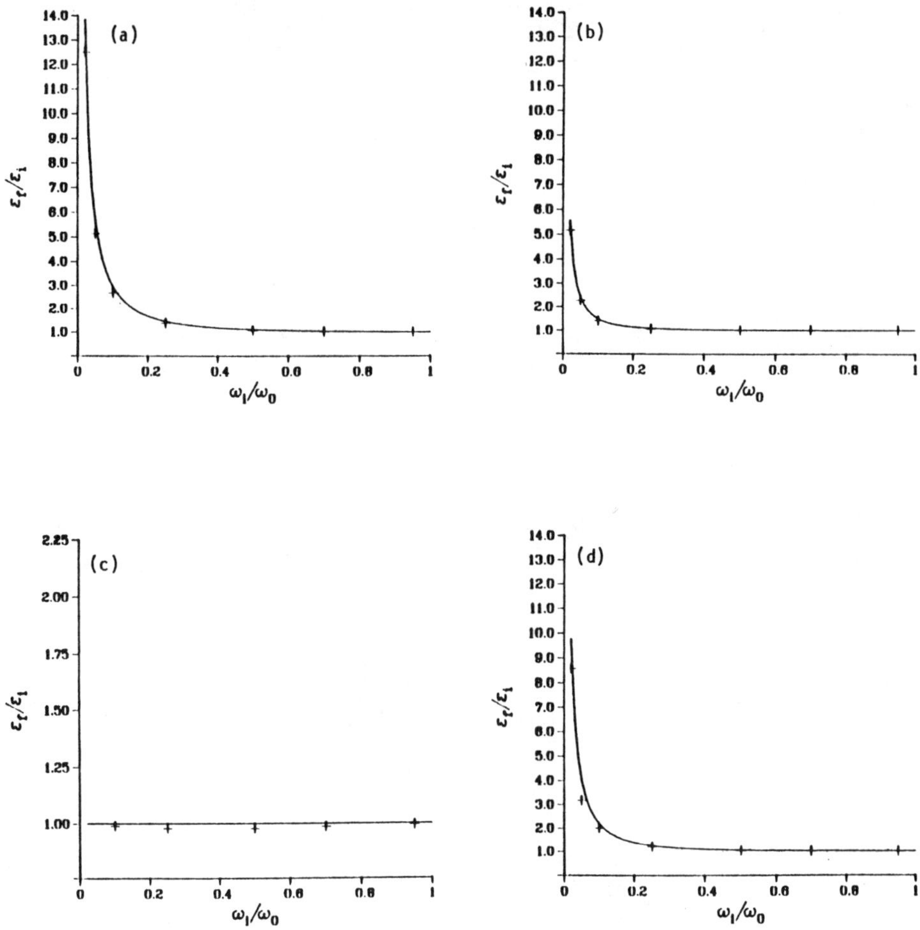

Fig. 8. Final emittance ratio versus initial tune ratio for initial (a) Gaussian, (b) waterbag, (c) thermal, and (d) hollow distributions. The curves are generated from the approximate formula, Eq. (14), using values of U/w_0 from Table 1, and the plus symbols show the results of the particle simulations after 100 plasma periods.

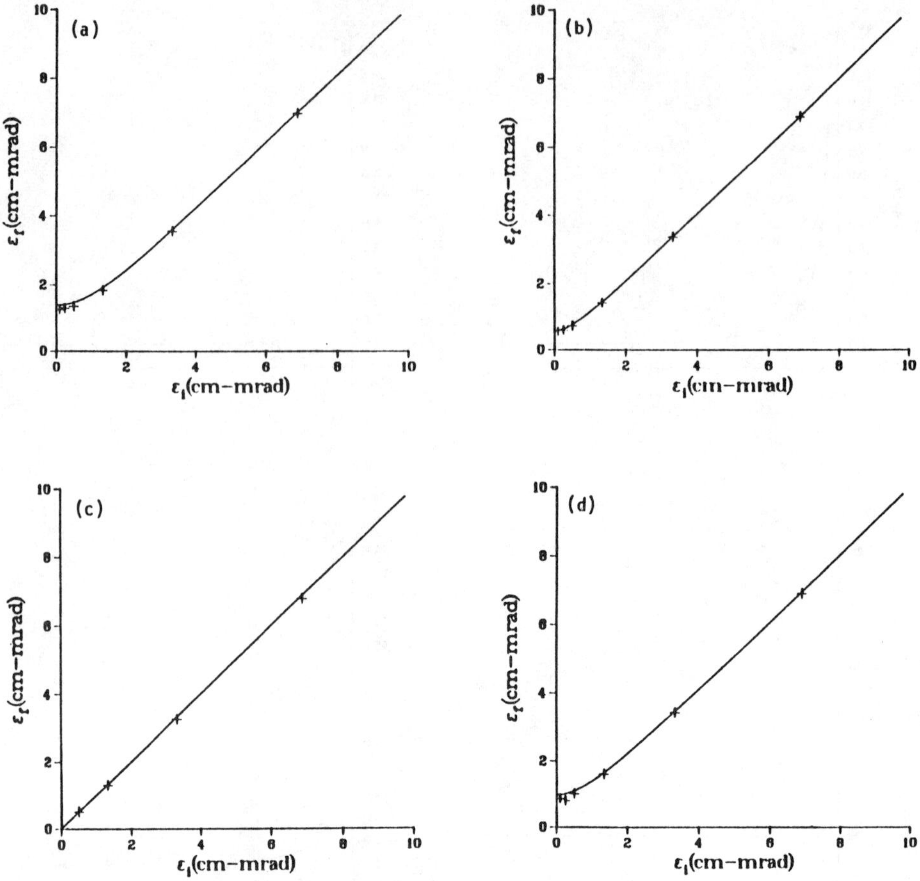

Fig. 9. Final emittance versus initial emittance for initial (a) Gaussian, (b) waterbag, (c) thermal, and (d) hollow distributions. The curves are generated from the approximate formula, Eq. (15), using values of U/w_0 from Table 1 and using a value $K\nu/\omega_0 = 5.0 \times 10^{-5}$ m. The plus symbols show the results of particle simulations after 100 plasma periods.

CONCLUSIONS

We have attempted to evaluate the ideas expressed by Struckmeier, Klabunde, and Reiser[8] that emittance growth is associated with the conversion of field energy to particle energy. We find that transverse energy conservation is valid for continuous channels, using the approximations given in this paper. We have obtained a formula, Eq. (10), showing that the rms emittance change can be related to three separate factors: rms beam size, perveance, and changes in a quantity we call the nonlinear field energy--a quantity that depends on the shape of the charge distribution and corresponds to the residual space-charge electric-field energy of beams with nonuniform charge distributions. When integrated for the case of an rms-matched beam, we obtained Eq. (13), which is in excellent agreement with our numerical simulation studies. Equations (10) and (13) relate changes in rms emittance to changes in the quantity U/w_0, regardless of the detailed mechanisms. The equations are applicable both for nonstationary beams and for stationary but unstable beams. Furthermore, when the final charge density is assumed to be uniform, the equations have led to good approximate formulas, including one that describes the well-known effect of minimum output emittance with decreasing input emittance.

The numerical studies reported in this paper have shown that the beams evolve to final states that are composed of a central core and a small halo, and that the emittance growth occurs within about one-quarter of a plasma period for high-intensity beams. From the numerical studies, we have seen examples of both the charge-density homogenization effects described in Ref. 8, resulting in an rms emittance increase, and of the reverse process, where the charge distribution evolves to one that is less uniform, resulting in a slight rms emittance decrease. The homogenization effect is a characteristic of peaked initial distributions in a highly space-charge-dominated case, whereas initially uniform distributions will become more nonuniform in a less space-charge-dominated situation, leading to a small rms emittance decrease. Equations (10) and (13) imply that undesirable rms emittance growth can be controlled by minimizing any decrease in the nonlinear field energy U, for example, by producing the beam with an initial uniform charge distribution. Then, field energy for rms emittance growth is unavailable. Both the semi-Gaussian (thermal) and the unstable K-V beams have provided good examples.

The studies in Ref. 8 were made for continuous beams in periodic quadrupole channels, and the results described in this paper apply for continuous beams in continuous focusing channels. Because continuous focusing channels represent a smooth-approximation representation of a periodic channel, we anticipate that our results will be approximately valid for periodic systems in a smooth approximation, if the betatron frequencies ω_0 and ω_i are replaced by phase advances per focusing period σ_0 and σ_i. We plan to conduct further numerical studies to test the formulas for periodic systems.

ACKNOWLEDGMENTS

We wish to thank J. D. Lawson, P. Lapostolle, R. H. Stokes, and M. Weiss for helpful communication and R. A. Jameson for his support. We acknowledge the diligent help of S. Watson in the preparation of this paper, and of S. O. Schriber for his valuable comments about the paper.

REFERENCES

1. R. Chasman, "Numerical Calculations of the Effects of Space-Charge on Six Dimensional Beam Dynamics in Proton Linear Accelerators," Proc. 1968 Proton Linear Accelerator Conference, Brookhaven National Laboratory report BNL 50120 (1968), 372.

2. R. A. Jameson and R. S. Mills, "On Emittance Growth in Linear Accelerators," Proc. 1979 Linear Accelerator Conf., Brookhaven National Laboratory report BNL 51134 (1979), 231.

3. L. Hofmann, L. J. Laslett, L. Smith, and I. Haber, "Stability of the Kapchinskii-Vladimirsii (K-V) Distribution in Long Periodic Transport Systems," Particle Accelerators 13 (1983), 145.

4. I. Hofmann and I. Bozsik, "Computer Simulation of Longitudinal-Transverse Space-Charge Effects in Bunched Beams," Proc. 1981 Linear Accelerator Conf., Los Alamos National Laboratory report LA-9234-C (1981), 116.

5. R. A. Jameson, "Equipartitioning in Linear Accelerators," Proc. 1981 Linear Accelerator Conf., Los Alamos National Laboratory report LA-9234-C (1981), 125.

6. R. A. Jameson, "Beam Intensity Limitations in Linear Accelerators," IEEE Trans. Nucl. Sci. 28 (3) (1981), 2408.

7. I. Hofmann, "Computer Simulation of High-Current Beam Transport," Proc. 1984 Linac Conf., Gesellschaft für Schwerionenforschung, Darmstadt report GSI-84-11 (September 1984), 304.

8. J. Struckmeier, J. Klabunde, and M. Reiser, "On the Stability and Emittance Growth of Different Particle Phase-Space Distributions in a Long Magnetic Quadrupole Channel," Particle Accelerators 15 (1984), 47.

9. R. Gluckstern, R. Chasman, and K. Crandall, "Stability of Phase Space Distributions in Two Dimensional Beams," Proc. 1970 Proton Linear Accelerator Conf., Batavia, Illinois (1970), 823.

10. P. M. Lapostolle, "Energy Relationships in Continuous Beams," Los Alamos National Laboratory translation LA-TR-80-8 or CERN-ISR-DI/71-6 (1971).

11. P. M. Lapostolle, "Possible Emittance Increase Through Filamentation Due to Space Charge in Continuous Beams," IEEE Trans. Nucl. Sci. 18 (3) (1971), 1101.

12. F. J. Sacherer, "RMS Envelope Equations With Space-Charge," IEEE Trans. Nucl. Sci. 18 (3) (1971), 1105.

13. J. D. Lawson, "The Physics of Charged Particle Beams," Clarendon Press, Oxford (1977), 197.

14. E. P. Lee, S. S. Yu, and W. A. Barletta, "Phase Space Distortion of a Heavy-Ion Beam Propagating Through a Vacuum Reactor Vessel," Nuclear Fusion 21 (1981), 961.

15. R. L. Gluckstern, "Oscillation Modes in Two Dimensional Beams," Proc. 1970 Proton Linear Accelerator Conf., Batavia, Illinois (1970), 811.

16. I. Hofmann, "Emittance Growth of Beams Close to the Space Charge Limit," IEEE Trans. Nucl. Sci. 28 (3)(1981), 2399.

SUMMARY OF 1981 HIGH-CURRENT BEAM DYNAMICS WORKSHOP

G.E. McMichael
Atomic Energy of Canada Limited, Research Company
Chalk River Nuclear Laboratories
Chalk River, Ontario, Canada KOJ 1J0

ABSTRACT

A workshop was held at the Chalk River Nuclear Laboratories October 28-30, 1981 to exchange information on beam dynamics problems of high current cw proton linacs, and to discuss new directions that might be followed. A general discussion of beam loss mechanisms and instabilities was followed by detailed discussions of particle beam dynamics codes and how these mechanisms are incorporated. Particular attention was given to current calculation capabilities and ways in which they could be improved. This paper gives a brief description of the topics that were discussed and lists the conclusions and suggestions for future work that were agreed upon in the summary sessions.

INTRODUCTION

The workshop, which was held in the week following the 1981 Santa Fe Linear Accelerator Conference, was informally structured. No formal papers were presented, although several participants gave informal talks on specific topics of current interest which provided a framework for discussions. All the sessions were tape recorded with the intention of publishing a lightly edited transcription. This was not done because the resulting transcription was too long and conversational. However, after the request for this review, we have done some additional smoothing to the proceedings to make them available by request. In this paper I will briefly describe some of the topics covered, the conclusions reached and further work suggested at that workshop, details of which can now be found in reference 1.

WORKSHOP AGENDA

The workshop was chaired by S.O. Schriber (CRNL). K.R. Crandall and T.P. Wangler (LANL), R.L. Gluckstern (U. of Maryland), M. Weiss (CERN) and K. Mittag (KfK), joined the CRNL participants B.G. Chidley, M.S. de Jong, E.A. Heighway, R.M. Hutcheon, G.E. McMichael and J. Ungrin for 2-1/2 days of informal lectures and round-table discussions. The topics covered in the various sessions were as follows:

1. Beam Loss Mechanisms

   K. Mittag listed the effects that he had considered as possible beam loss mechanisms for the SNQ study. This list formed the framework for discussions on the relative importance of different beam loss mechanisms, the effects that merit particular attention and the experimental tests that could be made to aid in understanding beam loss.

2. Beam Scrapers

This session dealt with the effect of scrapers on beam halo and in particular addressed the questions of whether halos can be removed with beam scrapers and if scrapers contribute to halo growth.

3. Beam Dynamics Codes

Talks by K. Mittag and K.R. Crandall describing the codes MOTION and PARMELA preceded a general discussion of the following topics.

(a) Space Charge Calculations
(b) Comparison of Calculations and the "Real World"
(c) Multi-Beamlet Beams
(d) PARMTEQ B and C
(e) Beam Dynamics Codes and Resonances
(f) Vane Shapes and Misalignments
(g) RFQ Design Recipes

4. Phase Space Reconstructions

R.L. Gluckstern described how to reconstruct the 4-dimensional phase space from its 2-dimensional projections. Extension to 6-dimensional space was also discussed.

5. Instabilities and Tunes

R.L. Gluckstern described how to integrate recent work on emittance growth and resonant instabilities (such as that reported by Hofmann and Bozsik[2]) into new designs and calculations.

6. Miscellaneous Topics

(a) Coherent and Incoherent Tunes (K. Mittag)
(b) Mismatches (M. Weiss)
(c) Field Measurements
(d) Accelerator Parameter Variation
(e) Halos and Scattering at Scrapers
(f) Other Beam Loss Mechanisms
(g) Errors

7. ZEBRA Accelerator Design

At the time of this workshop, the host Chalk River Nuclear Laboratories had just completed the beam dynamics design for a 300 mA, 10 MeV cw proton linac "ZEBRA". Details of this design were presented by B.G. Chidley and G.E. McMichael and triggered discussions on many aspects of high current RFQ and DTL beam dynamics.

8. Closing Session

The workshop topics were summarized and a list made of conclusions and suggestions for further work.

1981 WORKSHOP SUMMARY

The summary which follows, taken directly from reference 1, indicates the views held by the workshop participants of the then current state of high current beam dynamics and of some actions that could be taken to improve it.

A. Particle Beam Dynamics Codes

(a) Are useful for design purposes but predict performance, beam loss and emittance growth to probably no better than 5%.

(b) Cannot reliably calculate losses less than 1 part in 10^2 with any of the improvements being considered for the codes.

B. Small Beam Losses and How to Achieve Them

(a) Particle beam dynamics codes should be used to investigate mechanisms for loss and to study specific effects in some detail such as G. Boicourt at LANL is doing for halos.

(b) Real information about beam loss will have to come from experiments.

(c) Present operating experience should be surveyed, particularly regarding halos (e.g., LAMPF, FNAL, BNL, NEN, CERN). From the 1977 workshop, it appears that even if you clean up the beam, the halo soon returns.

(d) The mechanical beam hole radius should be no less than 5 to 7 times the beam rms radius (CERN criteria for safety factor to have small beam loss at high current).

(e) Optimum emittance should be used (not larger or smaller).

(f) Stay clear of instability regions.

C. Design Tools Presently Available

(a) Analytic techniques including such tools as CURLI, RFQUIK, "road maps" and ADAPT.

(b) Particle dynamics codes PARMxxx, MAPRO, MOTION.

D. Worthwhile Developments for Immediate Future

(a) RFQ Design - should look at other recipes as possible alternatives to those now used (e.g., variable V/r_0).

(b) ADAPT has been improved at KfK and a standard version should be adopted and distributed.

(c) MOTION (KfK) - It would be useful to put in RFQ fields, to consider a dc to bunched beam and to consider image charges.

(d) PARMTEQ, PARMILA, MAPRO
- Should put in the MOTION space charge routine for comparison and to look at multiple beamlets. Also, should look at lost particle handling algorithms, and how you shift fast or slow particles between buckets.
- Should improve the way the space charge is handled in the radial matching section of an RFQ.
- It will be difficult to put in image charges for an RFQ. Before this is done, should try to find simpler cases to

test how important image charges are or possibly look at image charges in DTL's.
- Should look at experimental measurements (e.g., POP) for fields in last cell of RFQ. Crandall's 3-D charge code may provide fields for PARMTEQ.
- May try more accurate boundary for computing particle loss in cells, but must also worry about the accuracy of the field representation (higher harmonics) in the region near the vanes.

(e) Analytic Tools
- Obtain road maps (Hofmann type diagrams) for wider range of $\varepsilon_x/\varepsilon_y$ values.
- Try to equipartition, but there may be other design constraints that are more important. In some cases, for example, you cannot equipartition and maintain longitudinal stability.
- Choose tunes to avoid instabilities.

E. Possible Sample Calculations to Estimate Seriousness of Effects

(a) RFQ image charge forces.
(b) Aperture aberration effects - e.g., solenoids, quads, etc.
(c) Errors, misalignments and transients.
(d) Scrapers (scattering probabilities possibly with Monte Carlo calculations).
(e) Scraper steering of a non-central beam.
(f) Multiple beamlets and characteristics.
(g) Halo studies using core as known.
(h) Check 150 mA current limit for CERN new 50 MeV linac (at approximately 2 MeV).

F. Miscellaneous Comments and Rules

(a) Avoid 60° and 90° phase advance per period.
(b) Avoid tunes where $n^*\sigma_t = m^*\sigma_l$.
(c) Good information on parameter space is required for the ion source and injector output (also changes due to oscillations or fluctuations).
(d) Fully neutralized and non-neutralized transport line benefits should be compared.
(e) Emittance growth reported for transport lines should be investigated. There is some worry here that the measuring device may affect what you are trying to measure.
(f) Must consider other mechanisms for losing particles.
(g) Multiparticle calculations and interpretations are limited by physical description of the system.
(h) It is advisable to clean up the beam bucket transversely (scrapers) and longitudinally (choppers and bunchers) as early as possible.
(i) Make changes smoothly, i.e., change accelerator parameters adiabatically.
(j) There is a need to understand the reasons for the different results from PARMTEQ-B and C.
(k) Use small gaps (< $\beta\ \lambda/2$) between tanks or plan to rebunch.

(l) RFQ designs for smaller output phase bucket should be investigated.
(m) Permanent magnet quadrupoles in drift tubes are a good idea but will limit the range of experiments.
(n) Be careful with scaling laws.
(o) Techniques for amplitude and phase control of tanks are a necessity.

ACKNOWLEDGMENTS

The contributions of the other editors and workshop participants are gratefully acknowledged.

CONCLUSIONS

Since the workshop in 1981, progress has been made on a lot of fronts. Experimentally, a number of laboratories now have operating RFQ accelerators or high current beam transport experiments to provide useful data to compare with calculations. I will not list the many code developments and new codes that are now available, except to point out that the RFQ image charge problem, which we had viewed as too difficult to tackle, is now being calculated both analytically[3] and using finite elements[4]. However, there still remain many questions to answer and the need for more and better tools and data. Beam spill and emittance growth continue to be hard beasts to tame and many of the fundamental problems of 3-1/2 years ago are still with us.

REFERENCES

1. B.G. Chidley, G.E. McMichael and M.S. de Jong. "Proceedings of the High Current Beam Dynamics Workshop". Unpublished report No. CRNL-2860, April 1985, Atomic Energy of Canada Limited, Research Company, Chalk River, Ont., K0J 1J0.
2. I. Hofmann and I. Bozsik, "Computer Simulation of Longitudinal-Transverse Space Charge Effects in Bunched Beams", Proc. 1981 Linac Conference, Los Alamos National Laboratory, Report LA-9234-C (1981) 116.
3. Yu-Yun Kuo, "Wall Effect in a Radio-Frequency Quadrupole", Nucl. Instr. and Meth. in Phys. Res., B10/11, 809 (1985).
4. B.G. Chidley and N.J. Diserens, "Beam Transmission of RFQ1 Calculated Using Finite Element Techniques for Space and Image Charges", Proceedings 1985 Particle Accelerator Conf., Vancouver, 1985 May 13-16 (to be published).

SUMMARY FOR WORKING GROUP ON NONINTERCEPTIVE DIAGNOSTICS

D. D. Chamberlin
Los Alamos National Laboratory, Los Alamos, New Mexico 87545

OVERVIEW

The working group on noninterceptive diagnostics spent much of its time comparing diagnostic techniques from different fields and their possible application to high power injectors. The group was composed of T. Fessenden, R. Fisher, D. Schneider, and D. Chamberlin and thus included backgrounds from electron machines, from fusion power diagnostics, from CW ion source and transport design, and from ion beam diagnostics. The probability of success for these possibilities is quite difficult to judge, short of a detailed examination of each application. Unexpected flaws or unforeseen noise problems can eventually vitiate an idea that at first sight appears promising. The following is a report of the several ideas that were discussed, with an indication of those most likely to succeed.

Wall current monitors, which measure the return currents in the beam pipe wall, offer both positional information from difference signals from separated current paths, and total current information from the sum. These have worked well in other applications.[1] The electronic components associated with electromagnetic pickups may be expected to perform well at frequencies up to a hundred megahertz, but both performance and component availability fall off rapidly as the frequency requirements increase toward and past one gigahertz. Similar information might be obtained from coupling a ferrite-loaded cavity to the beam pipe although this approach would work only for a bunched beam such as after an RFQ.

The optical sensing of beam-induced light is barely in the first stages of its applications. Several extensions immediately suggest themselves. The absolute level of light could provide more information about conditions in the beam region and could provide a useful monitor. It is possible that the electron flow in the DC portion of the beamline might create light and thereby be studied. In general, there is much more to gain from studying the physics of the excitation and light-production process. One could perturb the gas species and pressure. Information about ion velocity may be obtainable. Choosing specific lines with filters will give more specific data.

Beam potential measurements in the DC transport line are of great interest, not only for their relevance to ion source conditions and performance, but also in the study of neutralization of the beam space charge. The ideas included adding impurity background gas, photoionization, tracking of ion trajectories by a dye laser, and fast emissive probes. Electrons may be driven from the beam region with velocities depending on the potential. A further source of ideas and techniques is Reference 2.

A glaring lack in the present capabilities is in the detection of the beam halo. The optical technique may be able to see the beam edges by occluding the beam core with a light stop and then working hard to achieve sufficient dynamic range in the tails. The possibility of using a crossed beam diagnostic was noted. Also, the beam could be intercepted by a puff of gas or by a lithium curtain. It appears that the cross-sections are too low to observe any attenuation of an intercepting optical radiation, but perhaps a properly chosen electron beam could be used. A calculation of particle density shows the quite sparse ions of the beam to be passing through a much denser (by several orders of magnitude) background gas.

The service role of beam diagnostics was acknowledged by pointing out the usual evolution of a device from an experiment to a prototype and finally to a production tool. The importance of a measurement may vary in that the first reliable data may indicate proper conditions and the measurement may be dropped. But the diagnostic effort seems always to be called upon for more and more detailed information; thus, the effort must be organized to assure continuity of a team, and the continuing team must keep its publication records up to date.

Outlined below are the summary conclusions reached by the working group.

CURRENT TECHNOLOGY STATUS

- Interceptive techniques are well established for use on less intense beams.
- Noninterceptive optics techniques are working well for high-perveance, DC or CW beams in the measurement of beam profiles.
- Noninterceptive techniques are expected to evolve rapidly.
- Electromagnetic probes operate well at 80 MHz, but availability of electronic components will make the task more difficult at higher frequencies.
- Basically, the present needs are being met.

MAJOR PROBLEMS WITH CURRENT TECHNOLOGY

- The new work has been done by a small community.
- Interceptive techniques are generally not usable.
- High power density makes it vital to develop fast, high-quality diagnostics.
- Even beam halos are damaging to beamline components:
  - They are difficult to measure because they are dominated by the core regions, thus calling for a large dynamic range in the sensor.
  - They are difficult to simulate because their formation and behaviour is not understood and is poorly measured.
- There has been limited technology development for DC beams.
- Often the beamline design offers only limited space for equipment.
- Derived parameters require processing time.
- Absolute calibration of current is difficult without calorimetric methods.

IMPROVEMENTS TO PRESENT BEAM DIAGNOSTICS

There is a need for the following:

- Higher frequency response.
- Improved resolution.
- More real-time information.
- Extended dynamic range
  - Sensors that are sensitive to halo.
- Smaller sensors and instrument packages.
- Repackaged, better engineered components
  - Better reliability, data reproducibility.
- Enhanced (array) processing.

NEW IDEAS

- Beam potential measurements are of great interest for their impact on studies of space-charge neutralization.
- Absolute excitation light measurements
  - Electron flow may create light.
- Electron imaging.
- Current monitoring
  - Wall current.
  - Loaded cavity.

REFERENCES

1. T.J. Fessenden, B.W. Stallard, and G.G. Berg, ''Beam Current and Position Monitor for the Astron Accelerator,'' *Rev. Sci. Instru.* **43** *(12)*, 1789 (December 1972).
2. J. Borer and R. Jung, ''Diagnostics,'' CERN/LEP-BI/84/14 Geneva (October 1984).
3. D.D. Chamberlin, ''Noninterceptive Beam Diagnostics,'' in these Proceedings.

SUMMARY FOR WORKING GROUP ON THEORY AND SIMULATION FOR HIGH CURRENT R.F. ACCELERATORS

Lloyd Smith
Lawrence Berkeley Laboratory, Berkeley, California 94720

Because of the overlap in interest between this group and the High Current Beam Transport group, a significant fraction of the short time available was spent in listening to talks of common interest and in discussion thereof. The remainder of the time was spent in reviewing the status of specific R.F. linac simulation codes and formulating recommendations for future work. On these issues, McMichael's summary of the 1981 High Current Beam Dynamics Workshop provided a convenient framework for describing both the advances in code development since that time and recommendations for near-term action.

CODE DEVELOPMENT

1) PARMILA,[1] the oldest of the codes, and PARMTEQ, the RFQ version, has undergone no major changes since 1981.
2) PARMELA, a modification of PARMILA for electron accelerators, was developed in (about) 1981.
3) MAPRO, developed some years ago at CERN and spreading to other European laboratories, is an extension of PARMILA to full 3-D.
4) MOTION was developed at Karlsruhe, and treats space charge effects by direct inter-particle forces.
5) A still-unnamed code has been developed at Physical Dynamics[2,3] (Y.Y. Kuo) which also uses direct forces.
6) Another unnamed code was initiated at Chalk River and continued at Rutherford[4-6] (Diserens); it is 3-D (but four-fold symmetry) and includes image effects. Space charge is handled by a finite element method which has been used successfully in other fields of physics but is new to the electro-magnetic nature of R.F. linac simulation.

All these codes are operational, though the newer ones are still subject to accuracy tests and modification for efficiency in running time. It should also be pointed out that general purpose codes such as MASK and ARGUS are being applied to an increasing variety of problems and could, in principle, be used in R.F. linac work. Indeed, MASK was recently put to work in klystron design, a closely related field.

RECOMMENDATIONS

The contrast with McMichael's 1981 wish list is striking. We are well on the way to having computing capability for ''conventional'' linac problems, so that physics questions are not on the urgent list, with one exception. The mechanisms for, and consequences of neutralization processes have never been addressed except by the grossly oversimplified device of introducing a numerical factor in the self-force. The need for computational tools was recognized by the group, but to treat even the dynamics of ionization of residual gas, help will be needed from the other disciplines of physics. Situations in which significant numbers of electrons come from other sources (grids, wall, *et cetera*) may not be amenable to more than a phenomenological description.

The number of codes available is increasing and so is the number of potential users who are not in a position to write their own. Thus, communication is the problem of the day. Some specific suggestions are:

1) A repository of benchmark cases to compare the performance of codes with each other.
2) More standardized input-output.
3) On-line documentation; that is, to replace unreliable or non-existent narrative documentation by cards in the program which explain procedural questions and options.

There was also a plea for a grand package—a code which could take a designer from source to target with minimum specialized knowledge of codes or linac technology. This would require compatible inputs and outputs of the sub-codes at the very least and might well prove impossible by patching together existing codes which differ widely in physics assumptions and executions by incompatible computer systems.

The time may be approaching to form a committee or commission to grapple with these problems and try to establish some degree of standardization and effective channels of communication.

REFERENCES

1. D.A. Swenson, "Phase and Radial Motion in Linear Accelerators," TN-437 (Oct. 18, 1963).
2. Y. Kuo, "Beam Dynamics In Radio Frequency Quadrupole," IEEE Trans. Nucl. Sci. NS-28, 3464 (1981).
3. Y. Kuo, "Wall Effect on a Radio Frequency Quadrupole," Nucl. Instr. & Method in Phys. Res. B10/11, 809 (1985).
4. N.J. Diserens, "Investigation of the Effects of Space Charge and Image Charge Forces on Beam Loss in an RF Quadrupole," Proc. of the 1984 Linear Acc. Conf., GSI-84-11, 91 (1984).
5. N.J. Diserens, "Progress in the Development of a 3D Finite Element Computer Program to Calculate Space Image Charge Effects in RF Quadrupoles," Proc. of 1985 Particle Acc. Conf., to be published in IEEE Trans. Nucl. Sci. NS-32.
6. B.G. Chidley and N.J. Diserens "Beam Transmission of RFQ1 Calculated Using the Finite Element Method for Space and Image Charges," Proc. of 1985 Particle Acc. Conf., to be published in IEEE Trans. Nucl. Sci. NS-32.

SUMMARY FOR WORKING GROUP ON HIGH CURRENT BEAM TRANSPORT

Denis Keefe
Lawrence Berkeley Laboratory, Berkeley, California 94720

The high-current topics discussed in this working group can be categorized logically in the following order: (1) source, (2) injector, (3) choice of emittance definition, and (4) experimental results on high-current beam transport. Details on these topics are available in the references and invited papers in this Proceedings; only the briefest summary is presented here together with some personal observations.

1. SOURCE

The Oak Ridge group (Stirling, Whealton[1]) have applied a 3-D code to design the geometry of a surface-ionization H^- source and they predict that a current density of 8 amps/cm^2 should be obtainable in the exit beam, which will be a few millimeters across. The cesiated converter surface must be curved in both the x- and y-directions (and the transverse magnetic field surfaces maintained closely parallel to the curved converter); such a geometry results in a convergent gun in both x and y (convergence ratio in either plane is about eight). Encouragement that such a design might be achievable in practice comes from comparison of 2-D calculations of source performance, in which the emitter is curved in only one direction, with measurements on an operating source (SITEX). The beam size and current have been measured some distance away from the source exit and are *consistent* with emission conditions *at the converter surface* of 100 mA/cm^2 emission and a transverse temperature of about 0.1 eV. Because the beam is surely neutralized in propagating from the source, the size measurement does not *prove* such performance; instead an emittance measurement is needed, and this will take place in the next several months.

2. INJECTOR

Ultimately, one will seek for sources brighter than those available from plasma sources ($T > 1$ eV) or thermionic emitters/converters (< 0.1 eV). Maschke (TRW) stressed that in this endeavour, use could be made of the unique features of lasers with suitably chosen wavelength, because of their special ability to produce ions at a precise location, with exact timing.

When we finally come to achieve brighter sources, the scaling laws clearly point in the direction of a need for higher voltage injectors—2-3 MeV—perhaps with DC operation. Pierce column design becomes difficult at and beyond 1 MeV, the gradient on the insulator becoming uncomfortably large and the electrostatic focusing too feeble. Maschke urged more serious consideration of electrostatic-quadrupole focusing in a column (see Abramyan[2]) in which space-charge-limited currents can be transported (see Maschke[3]), but at an accelerating gradient as relaxed as one chooses.

3. EMITTANCE DEFINITION

Generally there is agreement that the varied definitions of emittance (normalized, unnormalized, RMS, $4 \times$ RMS, *et cetera*) are a cause for confusion and irritation. R. Keller (GSI) had still a further point to emphasize, namely, that it is often important to distinguish between the *circumscribed* emittance and the *4 × RMS* emittance.[4] He chose the special case of an S-shaped emittance, delivered, for example, by a source with spherical aberration, but small intrinsic temperature. In general, each of these definitions gives rise to an ellipse that is different in both area and orientation, and hence to different subsequent matching and aperature requirements. For the first choice, all particles are contained; for the second, some are inevitably lost. In the practical situation where one chooses to remove the most aberrant part of the beam—thus resulting in improved brightness with some small penalty in current—Keller argued that plotting the integral emittance as a function of beam fraction for the circumscribed ellipse gives one more insight into how best to make this judgement, rather than the more commonly used corresponding procedure for the $4 \times$ RMS ellipse.

4. HIGH-CURRENT BEAM TRANSPORT

Two experiments on the behavior of high-current heavy ion beams in a long quadrupole-focused transport system were widely discussed in this Working Group. These were:

(a) Magnetic quadrupole transport at GSI, discussed by Klabunde.[5] This experiment uses an argon beam (singly charged) at 190 KeV, and has a quadrupole triplet to accomplish matching. The transport line consists of six lattice periods (12 lenses) with two emittance-measuring stations, one at the entrance and the other at the exit. With the phase-advance per period set at 60 degrees, emittance growth was observed to occur when the depressed phase-advance became lower than 12 degrees, rising to a factor of two at 7 degrees. In a brief separate experiment, it was determined that this increase occurred very rapidly—after just one period, i.e., two lenses, into the transport section. Some time-dependence of emittance growth was observed which was due to neutralization by electrons.

(b) Electrostatic quadrupole transport at LBL, discussed by Kim[6] and by Keefe.[7] The beam was of singly-charged cesium at 120-200 KeV, five lenses were used for matching, and the FODO transport system was significantly longer than in the GSI case—41 periods (82 lenses)—with emittance-measuring stations at four locations. Empirically, beam transport was defined as stable if the current and emittance at the end of the system were found to be the same as the values measured at the entrance to the FODO lattice. No instabilities were detected for low-current phase advances less than 85 degrees—for all higher values, instabilities could be observed for high enough values of the beam current. In particular, for a low-current phase advance of 60 degrees, a depressed phase advance of 7 degrees was reached; no growth in emittance could be detected, in contrast with the GSI observation (see Fig. 1). Because electrostatic quadrupole lenses act also as clearing electrodes, no effects due to neutralization were observed.

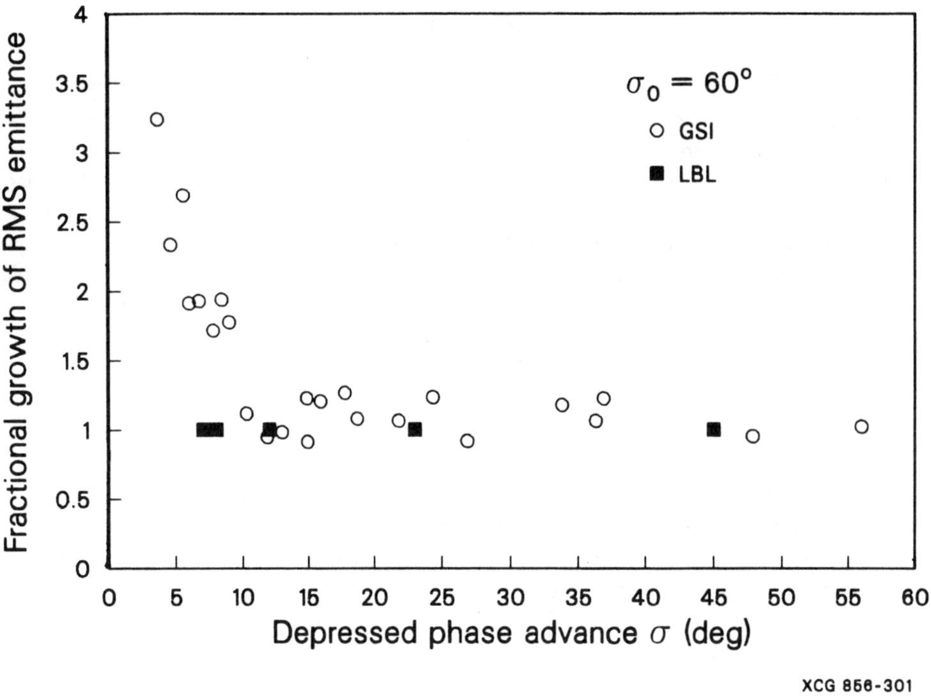

Figure 1. Emittance growth measured at GSI and LBL in a quadrupole transport system as a function of depressed phase-advance, σ. The GSI data show growth below $\sigma \sim 10$ deg, whereas the LBL group recorded no growth for the lowest value of σ reached (7 deg).

Happily, the discrepancy between these measurements admits of an easy explanation (qualitatively, at least) in terms of the simulation results presented by Wangler. The beam distribution in the GSI experiments was measured to be bell-shaped, and hence would be predicted to undergo rapid emittance growth as it relaxed to a nearly uniform one with consequent decrease in the non-linear field energy. The beam distribution function in the LBL case, however, was different, being well-approximated by a semi-Gaussian (uniform in space and Gaussian in velocity). Hence it would have a minimum nonlinear field energy to begin with, and should not develop emittance growth, from this effect at least. There are other reasons, of course, whereby the emittance could grow (see Celata *et alia*[8]); the important point to emphasize is that these were not observed in the LBL experiment, unless a significantly large displacement of a quadrupole was deliberately introduced. What part misalignments may have played in the GSI experiment still remains to be explored.

CONCLUSIONS

In recent years, new ground has been broken in several areas on the generation and handling of high-brightness space charge dominated beams at low energies. The push for ever brighter high-current sources is of crucial importance but, by itself, is not enough unless matched by improved capability to accelerate and transport still higher-current beams. The new results from LBL and GSI are extremely encouraging in that we now believe a linear quadrupole transport channel can transport much more current than thought possible a few years ago. It is important to continue these high-current transport experiments to find out where in fact the ultimate limits lie.

REFERENCES

1. J.H. Whealton, informal presentation at this Workshop.
2. E. Abramyan, *Trans. Nuc. Sci.*, NS-18, 447 (1971).
3. A.W. Maschke, Brookhaven Nat. Lab. Report No. 51692 (1983).
4. R. Keller, "High-Brightness, High-Current Ion Sources," in these Proceedings.
5. For a summary see M. Reiser, "High-Current Beam Transport and Charge-Neutralization Effects," in these Proceedings.
6. C.H. Kim, "High Current Beam Transport with Multiple Beam Arrays," in these Proceedings.
7. M. Tiefenback and D. Keefe, *IEEE Trans. Nuc. Sci.*, NS-32, (to be published, 1985).
8. C. Celata, I. Haber, L.J. Laslett, L. Smith, and M. Tiefenback, *ibidem.*

SUMMARY FOR WORKING GROUP ON CW RFQ DEVELOPMENT

S.O. Schriber
Los Alamos National Laboratory, Los Alamos, New Mexico 87545

The CW RFQ Development working group consisted of a good cross-section of the accelerator community with eight laboratories or institutions represented. Active participants were

Jerry Bueck, AFWL, Albuquerque, NM
Horst Klein, University of Frankfurt, Frankfurt, FDR
Peter Junior, University of Frankfurt, Frankfurt, FDR
Noburo Tokuda, INS, Tokyo, Japan
Will Stirling, ORNL, Oak Ridge, TN
Krsto Prelec, BNL, Long Island, NY
John Staples, LBL, Berkeley, CA
Gerry McMichael, CRNL, Chalk River, Ontario, Canada
Wayne Cornelius, LANL, Los Alamos, NM
Stan Schriber, LANL, Los Alamos, NM

Approximately nine hours of group interactions were used to discuss some of the aspects highlighted in the earlier review paper on CW RFQ development. Particular topics that were covered were RF penetrations, thermal conduction, thermal expansion and possible distortion, mechanical tolerances, RF windows, and materials selection.

RF PENETRATION

Measurements at CRNL have shown that penetration through the RF envelope should have a thickness to diameter or thickness to width ratio of at least 1.3 to reduce effective RF coupling to less than 1%.

BEAM DYNAMICS TOLERANCE

Dipole fields less than 5% of the quadrupole component will lead to acceptable RFQ performance. In general, dipole fields greater than 5% lead to degradation of performance—in transmission, emittance growth, and beam steering.

RF WINDOWS

The RF window has to be located carefully to avoid deleterious effects associated with X-rays or ultraviolet light emitted from the RF cavity volume.

Benefits of co-axial to rectangular geometrics need to be investigated more. More information exchange is required between the various groups to ensure an adequate understanding of all aspects of RF window design and location.

Coatings on the RF window surface have shown both improvement and degradation of performance. The reasons behind the change in performance must be clearly understood.

VANE-TO-VANE LONGITUDINAL JOINTS

Studies at INS have shown that making vanes out of a number of sections longitudinally does not create any significant effect at the vane joints.

RFQ MODELLING

Distributed network analysis such as that developed at ORNL should be improved to aid with the modelling of beam loading in the RFQ structure. The additions to existing computer codes should be straightforward leading to a means of investigating reactive and resistive beam loading, as well as in locations where it is most prevalent (the last part or acceleration portion of the RFQ).

RFQ GENERAL

RFQ technology is generally in a good state of development. Agreement between theory and experiment is relatively good. Mechanical engineering for a CW structure has demonstrated several suitable options with improvements as high-power accelerators are undergoing tests.

CRYOGENICS

Operation of an RF structure at 20° K should lead to reductions in structure losses by almost an order of magnitude. Under such improvements, one would have to investigate beam loading and its effects on overall operation. A well coordinated and adequately funded technology development program should be initiated.

WORKING GROUP RECOMMENDATIONS

The following recommendations were unamimously supported by the working group. The emphasis was clearly related to building a strong and viable technology base from which a number of similar and related projects could profit; both by information input and output. The impact of the developments in the near (up to two years), medium (about five years), and long (about ten years) term have also been identified for each of the recommendations.

1. Breakdown or Surface Characteristic Studies (near, medium, long)

Studies on phenomena that limit operating fields in RF structures should be strongly supported because of the many and varied potential payoffs. Collaborations already established are encouraged to continue with more interaction from metallurgists, materials scientists, surface physicists, and surface chemists. Studies should continue on effects of background gasses, surface geometry, surface metallurgy, and cryogenic applications. Work is underway at a number of laboratories in North

America, Europe, and Asia. A possible outcome from a collaborators meeting in September, 1985, at SLAC may be a more formal exchange of information, experience, and program plans.

2. High Power Testing Device (near, medium, long)

Developments in operating CW structures at high power certainly benefit from the existence of a dedicated RF facility with resonant load in which components and techniques can be tested at an early stage. Such activities have been very profitable at CRNL, University of Frankfurt, and DESY.

3. RFQ2 Project at CRNL (near, medium)

This 270 MHz RFQ for accelerating protons to 600 KeV is expected to have up to 75 mA of output CW beam by the end of 1986. An existing 400 KW CW RF source will drive the accelerator that will investigate beam parameter space, multiple beamlets, beam loadings up to 30%, multiple RF drives, and replacement vane geometry. The group at CRNL is encouraged to continue with their program and to interact with other groups operating CW devices.

4. Cryogenic Technology Base (medium, long)

Developments in this area could be extremely important for applications in the space environment. A coordinated and fully integrated program should be put into action with specific and reasonably attainable objectives. More interaction and discussion is required between interrelated groups to develop a much needed R&D plan.

5. TALL Project at INS (near, medium)

This 100 MHz RFQ accelerates heavy ions in a structure that is electrically long. INS is encouraged to continue because of the many implications and embellishments incorporated in this interesting project.

6. Geometry Studies at the University of Frankfurt (near, medium, long)

The group at the University of Frankfurt are encouraged to continue with their very productive R&D program. A number of RF sources spanning 10-200 MHz are available to assist with studies on electrodes, RFQ geometries (four rod, special four-rod, multi-rod, split coax four-rod, four vane), multiple beam arrays, surface breakdown, and general parametric investigations.

7. CW RFQ at LANL (very near, medium)

An 80 MHz 2 MeV RFQ is operating CW with 20 mA of protons routinely. Administrative control is limiting the current at present to be less than the design 100 mA. The facility could be used to test H^-, D^-, and H^+ beams as well as the present H_2^+ beam. The infrastructure built into the facility is invaluable and consists of diagnostics, control, RF and CW beam handling knowledge. Studies should continue on this important facility because of the obvious experience that users can gain with such a powerful CW beam.

8. Miscellaneous

A number of other items were discussed that could be useful for information available today and information that will be available in the future. The activities to be monitored include operations at GSI, CRNL, LANL, and DESY in CW activities other than those mentioned above, funneling studies, space charge neutralization, diagnostic development, fusion community developments, and the super HILAC operation.

SUMMARY FOR WORKING GROUP ON MULTIPLE BEAMS AND FUNNELING

Thomas P. Wangler
Los Alamos National Laboratory, Los Alamos, New Mexico 87545

The working group on Multiple Beams and Funneling discussed various topics related to multiple beams and funneling, including: (1) design considerations for multiple-beam accelerators; (2) scaling of current, emittance, and brightness for multiple-beam systems; (3) funneling lines using either discrete components or a radio-frequency quadrupole (RFQ) funneling structure; and (4) alternatives to funneling. The regular participants included C. Kim, G. Gillespie, R. Thomae, R. Stokes, G. McHarg, F. Guy, R. Burke, K. Bongardt, P. Krejcik, M. Shubaly, P. Grand, and T. Wangler.

MULTIPLE BEAM ARRAYS

In a practical beam transport or linac channel, the maximum beam current is limited by nonlinear contributions of the focussing elements, by the electric and magnetic focussing fields that are attainable, or by beam resonances or instabilities. To obtain higher currents than are possible in a single channel, multiple beams are required.

Maschke proposed the MEQALAC,[1] in which the beamlets are accelerated in common RF gaps and transported within individual channels using electrostatic quadrupoles. The concept was applied to low-velocity transport and acceleration. Maschke argued from linac current-limit formulae that the brightest linac beams could be obtained by filling the available aperture at the lowest practical velocity and highest possible RF frequency, implying the use of small apertures. The effective emittance, including all beams, is enlarged as a result of the geometrical dilution caused by the finite separation of the beams. Theoretically, it still might be possible to obtain a very bright beam in spite of this geometrical dilution effect by sufficient reduction of the bore size. Maschke and colleagues argued that it is feasible to construct electrostatic quadrupole arrays that contain *thousands* of separate channels,[1] resulting in a single high-current, low-emittance beam composed of thousands of individual beams. The first MEQALAC, the M1 MEQALAC,[1] was designed, built, and tested at Brookhaven in 1979. It accelerated nine beams of Xe^{+1} from 17 to 73 KeV and operated at a 4 MHz RF frequency. The bore diameters were 5/16 in. (0.79 cm). The measured average output current for nine beams was 2.8 mA, compared to a calculated 3.3 mA current limit.

The technical feasibility has not yet been demonstrated for producing a beam composed of thousands of small beams, within bore diameters that are orders of magnitude smaller than in the M1 MEQALAC. The MEQALAC is being investigated by the FOM Institute,[2] but this application seems primarily to satisfy a need for high current rather than high brightness. The concept proposed at FOM is a 16-channel, 100 mA, 6 MeV, Li^- MEQALAC for plasma diagnostics. In an initial study, they designed and constructed a MEQALAC to accelerate four He^+ beams from 40 to

0094-243X/86/1390173-5$3.00

120 KeV within a interdigital cavity resonator that operates at 40 MHz. The bore diameters are 0.6 cm. In the initial measurements, a beam current of 2.2 mA per beam was obtained a full energy, compared to a calculated current limit of about 3 mA. Although it is generally acknowledged that multichannel systems such as the MEQALAC potentially can produce high total beam current, there was no general agreement among the workshop participants that it is practical at the present time to drastically scale down the channel dimensions.

For some applications, such as heavy ion fusion, a high current density must be delivered to a target of radius r_T using a lens system whose effective usable radius r_L is limited because of aberrations and whose distance L from the target may be fixed. For this kind of application, the use of multiple-beam arrays may be especially well suited. If the lens is assumed to form a sufficiently small waist at the target, so that the waist size is determined primarily by the emittance term rather than the space-charge term in the envelope equation, then $r_T = L\varepsilon / r_L$ in the approximation that $2(L\varepsilon / r_T^2) >> 1$. Thus, the current density j at the target from a single beam can be written

$$j = \frac{r_L^2}{\pi L^2} \left[\frac{i}{\varepsilon^2} \right] ,$$

where i is the beam current. This result shows that a large current density at the target is obtained if the transverse brightness (i/ε^2) is large. However, space-charge-induced emittance growth, which occurs in a beam that is space-charge-dominated in the beam transport channel,[3] increases as a function of the quantity i/ε, so that although a large value of i/ε^2 is desirable for this application, a large value of i/ε is undesirable.

However, if the total current I at the target is composed of N overlapping beams of current i and emittance ε per beam, these apparently contradictory requirements may both be satisfied. We make the simple assumption[4] that $i \propto a^2$ and $\varepsilon \propto a$ for the beam extracted from an ion source, where a is the radius of the ion source extraction aperture, so that $\varepsilon \propto \sqrt{i}$. Then $i = I/N$, $\varepsilon \propto \sqrt{I/N}$, and $i/\varepsilon \propto \sqrt{I/N}$. The current density at the waist for the N overlapping beams is proportional to $I/\varepsilon^2 \propto N$. This argument suggests that if the total current at the target is provided by N independent channels, the current density at the target increases, and the quantity i/ε decreases as N increases. The geometrical dilution caused by the finite separation of the beams does not reduce the current density at the target.

The use of a multiple-beam array to deliver a high current density to a target is attractive for the induction-linac approach for heavy-ion fusion, which is being developed at the Lawrence Berkeley Laboratory (LBL). The MBE-4 induction linac at LBL[5] will model many of the features of such a multiple-beam system and will accelerate four space-charge-dominated beams from 0.2 to 0.8 MeV using individual bore diameters of 5.4 cm. The MBE-4 experiment will provide valuable information on acceleration, current amplification, beam handling and control, and emittance growth, which will be valuable for the design of future multiple-beam systems.

As discussed previously, the use of a multiple-beam accelerator is also necessary for applications that require higher currents than are possible in a single channel. An example is a concept for a high-power, multiple-beam accelerator for a 1 Ampere CW beam, as discussed at the workshop by Burke from the Hanford Engineering

Development Laboratory. The proposed uses of such an accelerator, include condensed-matter research, fusion-irradiation simulations, accelerator-breeder development, and strategic defense R&D. The accelerator system includes a multichannel RFQ and a multibore drift-tube linac using permanent magnets.

FUNNELING

The initial suggestion for funneling was made during the heavy-ion fusion workshops.[6] Because the space-charge limits relax with increasing energy, accelerator beams may reach an energy where the funneling of two high-current beams into one becomes attractive. In an RF linac, if the RF frequency is doubled, twice as many stable buckets per unit time are created, and the beam bunches from two linacs at frequency f can be funneled (radially combined and longitudinally interlaced), producing a beam suitable for injection into a new linac at frequency $2f$.

Funneling has several desirable consequences. It permits the use of more efficient high-frequency linac structures with all available buckets filled with particles, and it reduces the number of beams in a multiple-beam accelerator, resulting in reduced size, weight, complexity, and RF power. However, funneling may result in an unacceptable emittance growth, and the funneling line itself may contribute additional complexity to the overall accelerator system. For some applications, there may be alternatives to funneling, as was discussed at this workshop by Krejcik.[7]

The first simulation studies of beams in funneling lines were made by Bongardt,[8] whose results suggested that controlling emittance growth might be a difficult problem in some applications. Bongardt concluded that most of the emittance growth in his studies was caused by space-charge effects, together with some contribution from nonlinearities in the funneling-line elements. He explained the space-charge contribution as a result of an adaptation of the beam distribution to the funneling line, an explanation that has been supported by more recent studies.[9] Workshop discussions indicated that the requirements in the systems, studied by Bongardt for special elements (such as beam kickers that use up physical space), resulted in a reduced, effective, transverse-focussing strength in the funneling line. The reduced strength would be expected to cause an increased space-charge-induced emittance growth.

One of the challenges of a low-emittance-growth funneling line is to provide strong transverse focussing, especially while the two beams are being deflected by RF fields onto a common axis. Two solutions to this problem were presented during this workshop. One solution, presented by Guy,[10] made use of permanent-magnet quadrupoles, dipoles, and some combined-function elements to funnel two 2 MeV, 100 mA proton beams originating from a two-channel, 212.5 MHz RFQ. The overall funneling-line design was remarkably compact; it was contained within a total length of only 21.5 cm. A numerical simulation of the beam performance has yielded an emittance growth of only 15 percent. Further studies will be required to evaluate the potential technical difficulties that might be encountered with the combined function elements that were assumed in this design.

A second funneling approach was presented by Stokes[11] and uses a modified form of the RFQ to provide uninterrupted longitudinal and transverse focussing while funneling two beams. The RFQ funneling structure contains poles or vanes with a special periodic modulation pattern that results in sinusoidal time-varying quadrupole focussing, as in a conventional RFQ, but whose successive lenses are displaced alternately about the central axis in one transverse plane. This results in a time-dependent displacement of the neutral (zero restoring force) axis of the beam. Consequently, for two beams composed of bunches that are injected 180° out of phase, two neutral beam axes are created within the structure on opposite sides of the central axis. By a suitable programming of the modulation pattern, the two beams can be radially combined and interlaced into a single beam with twice the bunch frequency. Evaluations of the current limits and numerical simulation studies for beam performance have not yet been completed.

CONCLUSIONS

The early ideas on multiple beams and funneling that were stimulated by requirements for heavy-ion fusion accelerator systems have developed significantly. New experiments will provide important information necessary for a better understanding of multiple-beam systems. Important ideas have been suggested for funneling of high-current, low-emittance beams; however, experimental studies will be required for their evaluation.

REFERENCES

1. A.W. Maschke, "MEQALAC: A New Approach to Low Beta Acceleration," BNL 51029, (June 1979); and R.M. Mobley, *et alia*, "MEQALAC: A New Approach to Low Beta RF Acceleration," 1980 Conference on the Application of Accelerators in Research and Industry, *IEEE Trans. Nuc. Sci.*, NS-28, 1500 (April 1982).
2. R.W. Thomae, *et alia*, "The FOM-MEQALAC System: The RF accelerator," Proc. 1984 Linac Conf., Gesellschaft für Schwerionenforschung, Darmstadt report GSI-84-11, 118 (1984); and R.W. Thomae, F. Siebenlist, A. Schempp, F.G. Schonewille, E.H.A. Granneman, P.W. van Amersfoort, H. Klein, and T. Weiss, "The FOM-MEQALAC," in these Proceedings.
3. T.P. Wangler, K.R. Crandall, R.S. Mills, and M. Reiser, "Relation Between Field Energy and RMS Emittance in Intense Particle Beams," presented at the 1985 Particle Accelerator Conference, Vancouver, B.C., Canada, May 13-16, 1985; and T.P. Wangler, K.R. Crandall, R.S. Mills, and M. Reiser, "Field Energy and RMS Emittance in Intense Particle Beams," in these Proceedings.
4. M. Reiser, "Transport of High-Intensity Beams," presented at the 1985 Particle Accelerator Conference, Vancouver, B.C., Canada, May 13-16, 1985; and M. Reiser, "High-Current Beam Transport and Charge-Neutralization Effects," in these Proceedings.
5. R.T. Avery, C.S. Chavis, T.J. Fessenden, D.E. Gough, T.F. Hendersen, D. Keefe, J.R. Menenghetti, C.D. Pike, D.L. Vanecek, and A.I. Warwick, "MBE-4, A Heavy-Ion Multiple Beam Experiment," presented at the 1985 Particle Accelerator Conference, Vancouver, B.C., Canada, May 13-16, 1985; and C. Kim, "High Current Beam Transport with Multiple Beam Arrays," in these Proceedings.

6. D.E. Young, "Summary of RF Linac-Based System," *Proc. Heavy Ion Fusion Workshop*, Brookhaven Nat. Lab. Rept. BNL 50769, 17 (1977); also R. Kustom, A. Moretti, J. Staples, T.H. Khoe, and J. Keane, "Summary of Low-Beta Linac," *ibid.* 94; also D.A. Swenson, "RF Linac Approach to Heavy Ion Fusion," *Proc. Heavy Ion Fusion Workshop*, Lawrence Berkeley Lab. Rept. LBL-10301, 239 (1979).
7. P. Krejcik, "Optimising a High Current Injector to Incorporate a Fast Beam Kicker," presented at the 1985 Particle Accelerator Conference, Vancouver, B.C., Canada, May 13-16, 1985; and "A Cascaded RFQ Principle for High-Current Beams," in these Proceedings.
8. K. Bongardt and D. Sanitz, "Funneling of Heavy Ion Beams," *Proc. Symp. Accelerator Aspects of Heavy Ion Fusion*, Gesellschaft für Schwerionenforschung report GSI-82-8, 224 (1982); also K. Bongardt, "Emittance Increase in Heavy Ion Beam Funneling Lines," *Proc. 1984 Institute for Nuclear Study Intl. Symp. on Heavy Ion Accelerators and their Applications to Inertial Fusion*, 470 (Tokyo, 1984); also K. Bongardt, "Design Study of a High-Current Proton Beam Funneling Line," *Proc. 1984 Linac Conf.*, Gesellschaft für Schwerionenforschung, Darmstadt report GSI-84-11, 389 (1984).
9. F.W. Guy and T.P. Wangler, "Funneling: An Initial Beam Dynamics Study," in these Proceedings.
10. F.W. Guy, "A Low Emittance Growth Funneling Line: Preliminary Design and Beam Dynamics Study," in these Proceedings.
11. R.H. Stokes and G.N. Minerbo, "Proposed Use of the Radio-Frequency Quadrupole Structure to Funnel High Current Ion Beams," presented at the 1985 Particle Accelerator Conference, Vancouver, B.C., Canada, May 13-16, 1985; and R.H. Stokes and G.N. Minerbo, "Scheme to Funnel Ion Beams with a Radio-Frequency Quadrupole," in these Proceedings.

A CASCADED RFQ PRINCIPLE FOR HIGH-CURRENT BEAMS

Patrick Krejcik
SNQ-ABT, Kernforschungsanlage Jülich,
5170 Jülich, Fed. Rep. Germany.

Introduction

In this paper a method is discussed for increasing the current limit in an injector linac using RFQ's. The method is based on making a frequency jump between two RFQ's by allowing the beam to debunch after the first RFQ and rebunching the dc beam at a higher frequency in the second RFQ. A frequency jump is usually necessary in an injector chain in order to satisfy all the design boundary conditions at the required beam current. These boundary conditions appear in a simple design formula for the current limit in a RFQ, namely

$$I \propto (E_m/E_k)^2 U_o^{1/2}/f \tag{1}$$

The maximum electric field, E_m, in the RFQ is chosen with respect to the Kilpatrick field level, E_k, to be as high as possible without degrading the reliability of the RFQ with excessive sparking. The injection energy into the RFQ is limited by the practical upper boundary on the ion source voltage, U_o. Thus in order to be able to accelerate very high currents an operating frequency, f, has to be chosen for the first accelerator which is lower than the optimum frequency for the high energy accelerating stages.

A further point to consider is that an injector must usually incorporate a large number of diagnostic devices and subsidiary systems such as fast beam kickers. The design must therefore be optimized to allow space for these systems as well[1].

In making the frequency jump to the high energy stage it is also desirable that all the high frequency rf buckets are filled so that efficient use is made of the rf power in the high energy accelerator. This can be done in general with a funneling system, for which several schemes have been discussed at this workshop. However, at low energies it is also possible to make use of the special properties of RFQ accelerators to make an arbitrary frequency transition.

Debunching and Capture by RFQ's

The good bunching characteristics of RFQ's combined with transmission efficiencies in excess of 90% make it possible to consider schemes in which the capture of a dc beam and bunching is performed more than once in the linac. For example, the low energy beam from the ion source can be accelerated in a RFQ whose frequency is chosen low enough to allow the full current to be accelerated. If the beam is accelerated to an intermediate energy and then debunched into a dc beam once more it can be captured in a second RFQ and rebunched at an arbitrarily higher frequency. This higher frequency is chosen to match the rf of the high

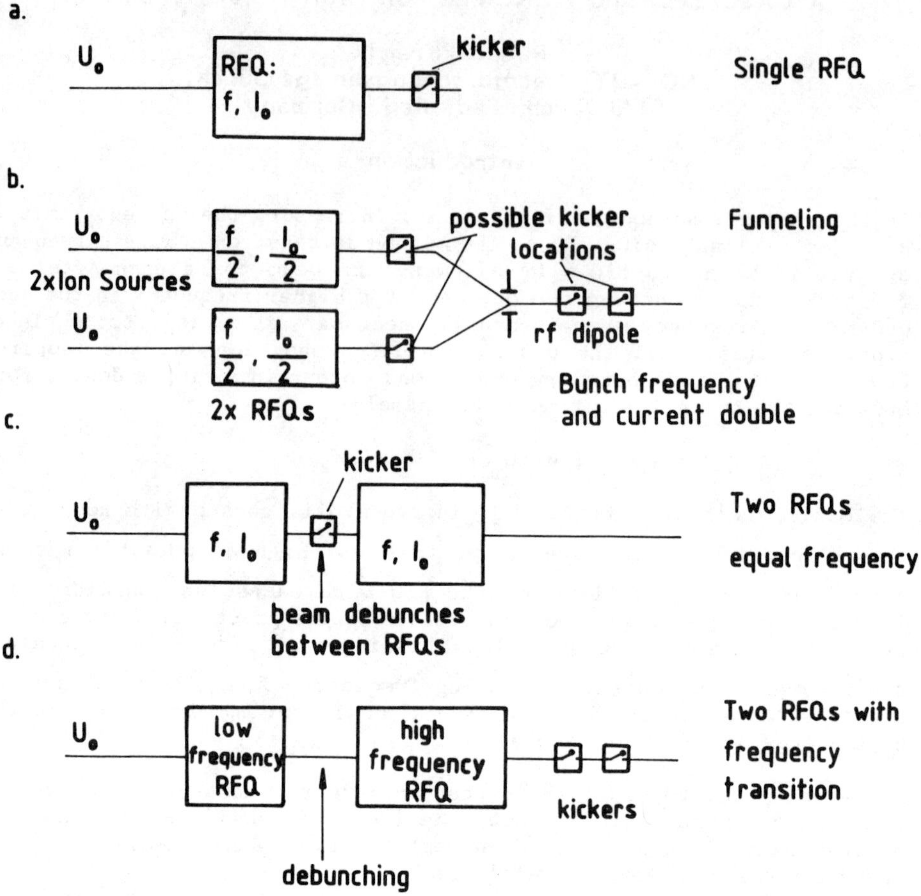

Fig. 1. Comparison of high-current injector configurations where the boundary conditions are given by the ion source voltage, U_o, and the final energy, U_f, and frequency, f, of the beam. Position of a fast bunch kicker is also shown.

a. single RFQ whose current is determined by the above constraints.
b. funneling, in order to double the bunch frequency and current.
c. two RFQ's operating at the same frequency in order to incorporate the kicker at a lower energy.
d. two RFQ's with an arbitrary large frequency transition to achieve a higher current limit.

energy accelerator. The energy at which the beam is injected into the second RFQ should thus be chosen high enough so that the current limit remains the same as in the first RFQ. In general, the increase in frequency that is allowable from one RFQ to the next is proportional to the increase in the particle's velocity.

To illustrate this it is possible to consider the parameters from the proposed German spallation neutron source[2] (SNQ) where 200 mA of protons are to be extracted from the ion source at 50 keV and accelerated to 2 MeV for injection at 200 MHz into the next stage. Rather than doing this with a single 200 MHz RFQ, where the maximum electric fields may be too high, it is possible to consider either funneling two 100 MHz, 100 mA beams, or a cascaded 200 mA RFQ scheme. In a cascaded RFQ scheme the 200 mA beam could be accelerated from 50 keV to 200 keV by a 100 MHz RFQ and then debunched to be recaptured and bunched at 200 MHz in a second RFQ for acceleration to 2 MeV.

Care must be taken with the debunching of the beam at the intermediate energy to avoid making a longitudinal mismatch resulting in an emittance dilution. If a bunched beam is allowed to debunch while drifting, as in fig. 2, then the effective emittance of the dc beam is 2π times the energy spread, which can be much larger than the area of the longitudinal phase ellipse. Proper matching of the beam requires that the longitudinal phase ellipse be rotated to give minimum energy spread.

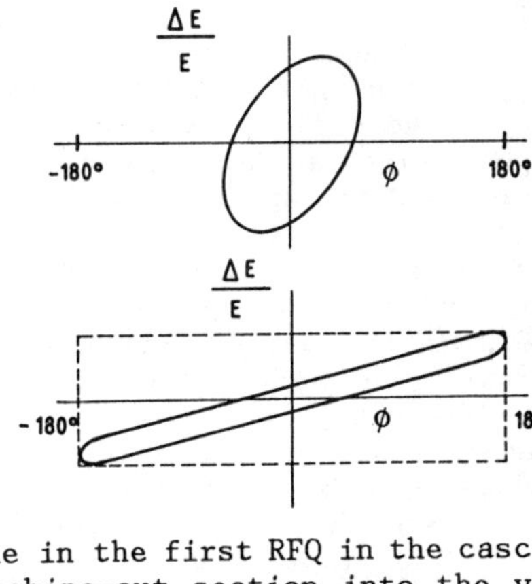

Fig. 2. The longitudinal emittance of a bunched beam represented by an ellipse (top) is diluted to the rectangular area (bottom) when allowed to debunch in a drift space to a dc beam.

This longitudinal matching can be done in the first RFQ in the cascade by incorporating a debuncher and matching-out section into the vane profile at the end of the RFQ. Such a procedure has already been successfully tried in the Saclay RFQ[3], where a dc beam was required for other reasons.

The frequency transition in going from one RFQ to the next should not produce any intensity modulations in the beam, so it is important that the rf structure in the beam from the first RFQ be completely smeared out. Preliminary studies[4] with PARMTEQ have shown that the initial gaussian intensity distribution in the bunch is smeared out during debunching as a result of the strongly filamented structure present in the bunches.

RFQ Design Considerations

The cascade configuration is able to deliver more current than a single RFQ since more free parameters are available to satisfy all of the boundary conditions. The current limit in the first RFQ is made high by choosing a low frequency to match the low injection voltage of the ion source. The current limit in the second RFQ is made high by choosing an injection energy that is high enough to match the high operating frequency of the high energy accelerator.

The higher injection energy into the second RFQ influences the length of its gentle bunching section and hence the length of the RFQ. The gentle buncher is made up of a number of $\beta\lambda$ cells so that at higher β the length of the RFQ measured in units of λ becomes greater. In a four vane resonator this makes the problem of achieving voltage flatness a little more difficult, but this has been overcome in existing RFQ's. The physical length of the $\beta\lambda$ cell does not increase in the second RFQ because λ is correspondingly smaller. However, the number of $\beta\lambda$ cells required for the gentle buncher may not be the same at different frequencies so it is difficult to say in general if the physical length increases greatly with the higher injection energy.

The first RFQ, on the other hand, is fairly short because of its low end energy. This, plus its large transverse dimensions due to low frequency, would allow it to be built more robustly to tolerate, for example, a longer duty cycle or higher beam losses in comparison to a single high-energy RFQ. The low end energy also implies that the beam is not very strongly bunched in the RFQ. Thus the current limit is higher in this short RFQ than would be expected for a single, high-energy RFQ operating with the same injection energy and frequency.

The increase in current limit resulting from the low end energy of the first RFQ allows another configuration to be considered where the two RFQ's in the cascade operate at the same frequency. If the beam only partially debunches at the intermediate energy and is recaptured at the same frequency in the second RFQ, it is possible to incorporate a fast beam chopper at the intermediate energy stage. This makes the beam chopper system shorter than if it were incorporated at the end of the high-energy RFQ and obviates the need for rebunchers in the chopper beam line since rebunching is performed adiabatically in the second RFQ. The increase in current limit over a single RFQ is not as dramatic as when a frequency transition is made, but it may none the less find application in some circumstances.

Conclusion

Debunching of the beam in a cascaded RFQ scheme is an alternative method to multiple beams and funneling as a means of increasing the current limit in an accelerator. The final beam brightness is often of equal importance as the total current that can be accelerated so it is also necessary to consider all factors that contribute to the final emittance of the beam. The importance of correct longitudinal matching during debunching has been discussed for the cascaded scheme, apart from which normal RFQ design procedures would be applicable to minimize longitudinal and transverse emittance growth. In a funneling scheme care must be taken with the layout of the funneling beam lines, particularly the aperture limits of the rebuncher cavities, so that emittance growth is also not excessive.

In the cascaded RFQ system the onus of delivering the full current falls back once more to the ion source and low energy beam transport system. As a first approximation, if the source brightness is assumed constant, the emittance of the beam would be twice that of the individual beams in an equivalent funneling scheme. The demand for the full current to be delivered by a single low energy beam transport system is offset to some extent by the ability to design the first RFQ in the cascade more robustly which in turn allows some of design restraints in the low energy beam line to be relaxed.

In addition to the low emittance requirements, some applications also require that there be minimal intensity fluctuations in the beam. In the cascaded scheme the debunching has to be complete enough to prevent any intensity modulations. This effect has to be weighed against systems that use more than one beam line, where the characteristics of the two ion sources, for example, can drift apart during operation and produce an intensity ripple in the combined beam.

References

1. P. Krejcik, Proc. 1985 Part. Acc. Conf., Vancouver, May 1985.
2. C. Zettler, Proc. 1984 Lin. Acc. Conf., GSI 84-11, pp. 480-484.
3. J.L. Laclare and A. Ropert, LNS Getis 063, Saclay 1982.
4. P. Krejcik and R. Lehman, int. report, unpubl.

FUNNELING: AN INITIAL BEAM-DYNAMICS STUDY*

F. W. Guy and T. P. Wangler, AT-1, MS H817
Los Alamos National Laboratory, Los Alamos, NM 87545

ABSTRACT

Funneling two H^- beams into a single beam of twice the current has been examined as a means of doubling beam current without significantly increasing transverse emittance. Using the PARMILA particle-following code, two 100-mA RFQ output beams at 2 MeV were injected into idealized transport lines for merging two beams into one. Two approaches were studied: (1) the "minimum-element" method, in which a minimum number of discrete elements such as quadrupole triplets, buncher cavities, and bending magnets were used to transport and deflect the beam; and (2) the "quasi-adiabatic" method, in which a periodic lattice similar to the RFQ provided focusing and minimized abrupt changes in the beam environment. The minimum-element method resulted in an emittance growth ratio $\varepsilon_0/\varepsilon_i = 2.5$, whereas the quasi-adiabatic emittance growth ratio was about 1.1 (albeit with an idealized line configuration).

INTRODUCTION

In its simplest form, funneling combines two beams of frequency f and current I by longitudinal interlacing of bunches into a single beam of current 2I, suitable for injection into a linac of frequency 2f. Funneling of high-current beams in rf-linac systems can be advantageous when the current limits, which increase with increasing energy and decrease with increasing frequency, allow a frequency doubling for subsequent accelerating sections. After frequency doubling, the number of available buckets per unit time doubles, and all the buckets can be filled by funneling. A funneling line might contain bending magnets to bend the beam, quadrupole lenses, and rebuncher cavities to focus the beam radially and longitudinally, and rf electric deflectors that produce time-varying transverse electric fields to merge the beams. Ideally, funneling would double the beam current per channel with no increase in transverse emittance. Funneling can be cost effective because it reduces the number of linac lines. However, in practice some emittance growth will occur because of space-charge nonlinearities and because of nonlinear, chromatic, and time-dependent effects associated with the funneling-line elements. Our objective in this study is to understand these effects and to minimize this growth.

The initial suggestion for funneling was made at the 1977 Heavy Ion Fusion Workshop.[1,2] The first simulation studies of beams in funneling lines were made by Bongardt.[3-5] Of special interest was

*Work supported by US Dept. of Defense and the Ballistic Missile Advanced Technology Center.

Bongardt's study of the funneling of two 100-mA H^+ beams at 2 MeV for the German neutron spallation source project.[5] Bongardt reported an emittance growth ratio (output emittance divided by input emittance) of $\varepsilon_o/\varepsilon_i = 2.4$ for an initial normalized rms emittance $\varepsilon_i = 0.06$ π•cm•mrad, a result suggesting that controlling emittance growth might be difficult. Bongardt concluded that the calculated emittance growth was caused partly by space charge and partly by nonlinearities in the funneling-line elements. He explained the space-charge contribution as a result of the adaptation of the beam distribution to the funneling line.

APPROACH

In our funneling investigation, we chose to begin by studying funneling of two 100-mA H^- beams at 2 MeV from a pair of 212.5-MHz RFQ lines for injection into a 200-mA, 425-MHz drift-tube linac. Other energies could have been studied, but 2 MeV was chosen because it is a typical final energy for the RFQ. An RFQ design study for this problem resulted in a design with an output emittance slightly greater than 0.02 π•cm•mrad.

For numerical simulations and for emittance growth calculations, we used the particle-following program PARMILA with both 2-D and 3-D space-charge treatments. Simulation runs were made with about 300 particles, starting with the output RFQ particle coordinates taken from an earlier RFQ simulation. Runs were also made with 500 particles using the RFQ output-beam ellipse parameters but with uniform spatial distribution. Idealized hard-edged elements were assumed for the funneling line and only on-axis beams were studied. Magnet pole-tip fields were restricted to $B \leq 12$ kG and rf electric fields were restricted to $E \leq 1.5\ E_K$, where E_K is the usual Kilpatrick limit. Quadrupole radial apertures were about twice the beam radius. The interaction of the two merging beams has not yet been taken into account; thus, we began by studying emittance growth in a single beam.

The requirement of minimizing emittance growth leads to the following general conclusions. The radial and longitudinal beam size should be kept small in the rebuncher cavities to minimize nonlinearities. The beam size should be kept small in the deflecting plane of bending magnets and rf deflectors to reduce chromatic contributions. The longitudinal beam size should be kept small in the rf deflector to minimize emittance increase from the time-dependent deflection. Strong focusing should be provided to prevent a large space-charge tune depression. Finally, from our experience in designing RFQs, we concluded that it is desirable to minimize abrupt large changes in the charge distribution, leading to space-charge-induced emittance growth.

We have investigated two rather different approaches to the design of funneling lines. These we call the minimum-element method and the quasi-adiabatic method. In the minimum-element method, we tried to use the minimum number of conventional elements to transport and deflect the beam. Large individual bending angles were used to reduce the number of elements and the overall length. With this approach, we did not worry about abrupt changes affecting the charge distribution. In the quasi-adiabatic method, we tried to minimize changes in the charge distribution. Focusing and bunching were performed by a periodic lattice of the same frequency and strength as the RFQ lattice, and the total bend angle was kept small and distributed over several elements.

CALCULATIONAL RESULTS

Preliminary results are given in Table I. Beam envelopes, emittances, and line diagrams are shown in Figs. 1 and 2. In the figures, z-emittances have been expressed in the same units (π•cm•mrad) as transverse emittances. Details of beamline elements are shown in Tables II and III. The minimum-element method, tested for two single-channel RFQs, resulted in an emittance growth ratio of 2.5, a result similar to that of Bongardt.[5] The quasi-adiabatic method was tested for a double-channel RFQ with a small initial beam separation. A double-channel RFQ incorporates an rf structure that focuses and accelerates two separate beams. This method yielded the very encouraging result of $\varepsilon_0/\varepsilon_i = 1.1$. However, the quasi-adiabatic method was studied for a highly idealized funneling-line configuration that did not allow for merging of the two beams in a practical way. We found that emittance growth for both methods was caused primarily by space charge. The two methods were applied to different problems in that different initial beam separations were used, so an exact comparison between the methods is not strictly valid. However, little of the growth in the minimum-element method was due to the bending elements, so that even if this method could be used with the double-channel RFQ, reduction of the bend angle might not greatly change the results.

Table I. Preliminary emittance growth results

Method	RFQ Configuration	Emittance Growth $\varepsilon_0/\varepsilon_i$
Minimum element	two single-channel RFQs	2.5
Quasi-adiabatic	one double-channel RFQ	1.1

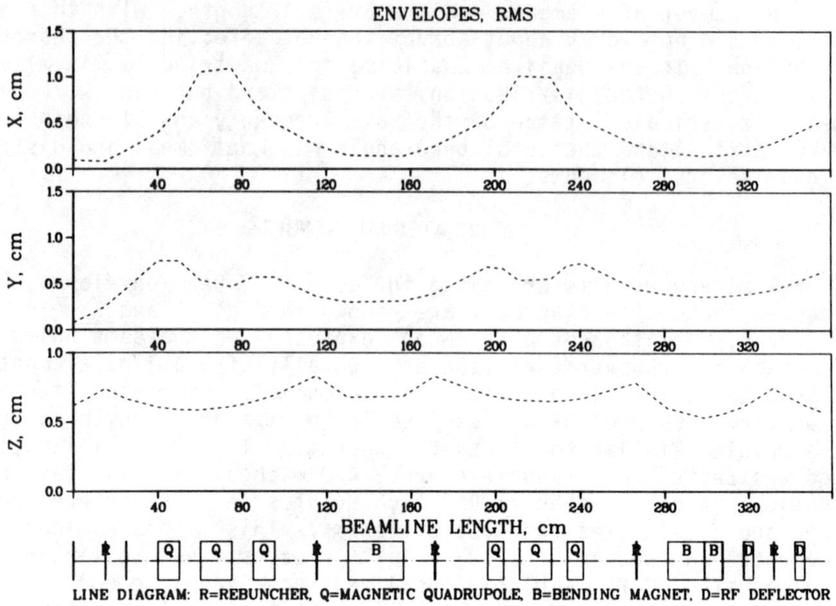

Fig. 1A. Minimum-element line envelopes.

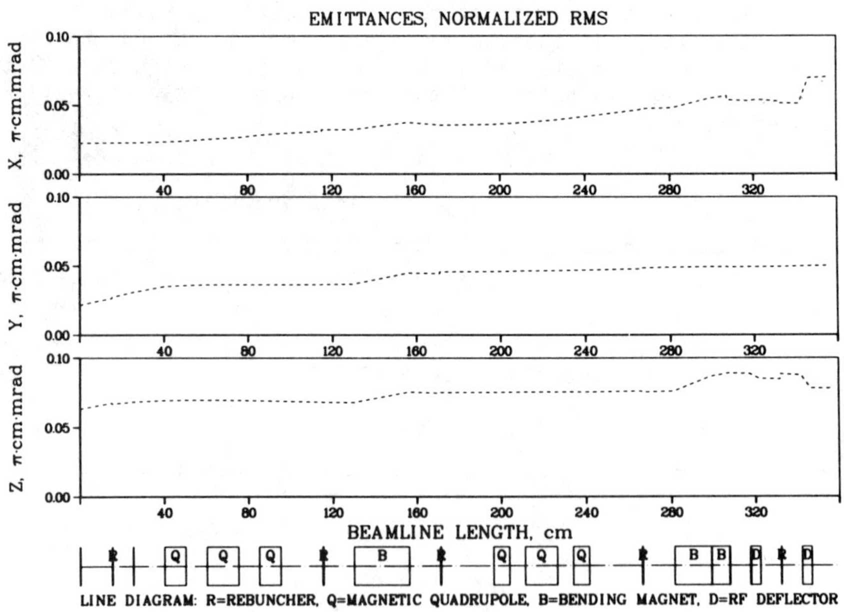

Fig. 1B. Minimum-element line emittances.

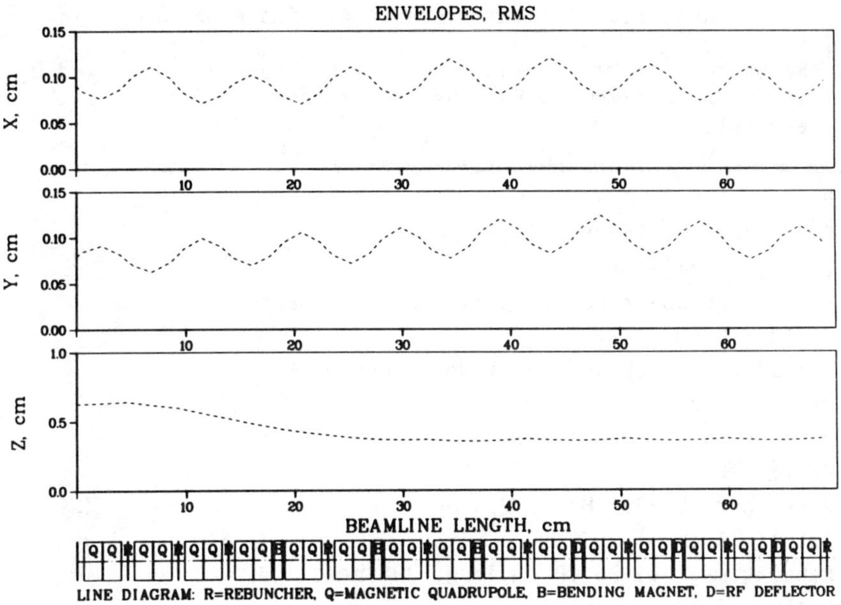

Fig. 2A. Quasi-adiabatic line envelopes.

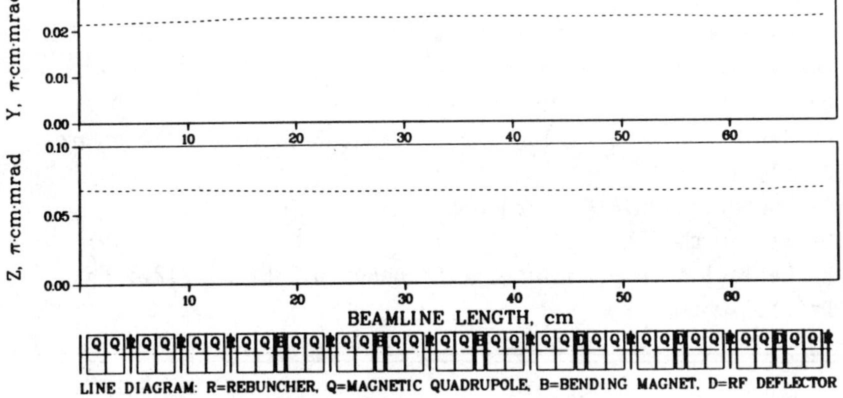

Fig. 2B. Quasi-adiabatic line emittances.

Table II. Minimum-element beamline elements

Note: Rebunchers are zero length but are preceded and followed by drift space allowing for the actual physical length of the element.

1. Drift, 15 cm
2. Rebuncher, 0.1754 MV, 212.5 MHz
3. Drift, 25 cm
4. Defocusing quadrupole, 10 cm long, 1100 G/cm
5. Drift, 10 cm
6. Focusing quadrupole, 15 cm long, 895 G/cm
7. Drift, 10 cm
8. Defocusing quadrupole, 10 cm long, 760 G/cm
9. Drift, 20 cm
10. Rebuncher, 0.1500 MV, 212.5 MHz
11. Drift, 15 cm
12. Bending magnet, 26.18 cm long, 30° bend
13. Drift, 15 cm
14. Rebuncher, 0.1369 MV, 212.5 MHz
15. Drift, 25 cm
16. Defocusing quadrupole, 7.5 cm long, 1160 G/cm
17. Drift, 7.5 cm
18. Focusing quadrupole, 15 cm long, 950 G/cm
19. Drift, 7.5 cm
20. Defocusing quadrupole, 7.5 cm long, 1160 G/cm
21. Drift, 25 cm
22. Rebuncher, 0.1565 MV, 212.5 MHz
23. Drift, 15 cm
24. Bending magnet, 17.45 cm long, 20° bend
25. Septum magnet, 8.73 cm long, 5°bend
26. Drift, 10 cm
27. RF deflector, 4.6 cm long, 2.5° bend, 6.15 MV/m, 212.5 MHz
28. Drift, 10 cm
29. Rebuncher, 0.0874 MV, 425 MHz
30. Drift, 10 cm
31. RF Deflector, 4.6 cm long, 2.5° bend, 6.1 MV/m, 212.5 MHz
32. Drift, 10 cm

Table III. Quasi-adiabatic beamline elements

Note: See Fig. 2A and B for element arrangement. Elements are combined in modules of 4.6 cm length, corresponding to $\beta\lambda$ for a 2-MeV H^- particle at 425 MHz.

1. Permanent-magnet quadrupoles: 3.4 cm long, 14500 G/cm to 16500 G/cm
2. Rebunchers, 1.2 cm long, 0.02 to 0.11 MV, 425 MHz
3. Permanent-magnet bending magnets, 1.2 cm long, 1.75° bend
4. RF Deflector, 1.2 cm long, 1.75° bend, 10.5 MV/m, 212.5 MHz

CONCLUSIONS

We have calculated two types of funneling lines, the minimum-element line and the quasi-adiabatic line. For applications in which funneling must be employed for the purpose of doubling beam current with the least possible growth in emittance, we have concluded that the quasi-adiabatic line shows great promise. Additional work is necessary to confirm these preliminary results and to continue this study with more realistic elements and greater detail in our computer models. We are investigating ideas for special elements such as combined function magnets and two-beam rf structures; we need clever ideas for methods to maintain tight focusing in both the transverse and longitudinal directions and to accomplish beam merging at the same time.

ACKNOWLEDGMENTS

We wish to thank J. M. Potter and R. H. Stokes for many helpful discussions, and K. R. Crandall and R. S. Mills for their help on beamline design and assistance with PARMILA.

REFERENCES

1. D. E. Young, Proc. Heavy Ion Fusion Workshop, Brookhaven National Laboratory report BNL- 50769 (1977), 17.
2. R. Kustom, A. Moretti, J. Staples, T. H. Khoe, and J. Keane, ibid., 94.
3. K. Bongardt and D. Sanitz, Proc. Symp. Accelerator Aspects of Heavy Ion Fusion, Gesellschaft für Schwerionenforschung report GSI-82-8 (1982), 224.
4. K. Bongardt, Proc. of 1984 Institute for Nuclear Study International Symposium on Heavy Ion Accelerators and Their Applications to Inertial Fusion, Tokyo, Japan (1984), 470.
5. K. Bongardt, Proc. 1984 Linac Conf., Gesellschaft für Schwerionenforschung, Darmstadt report GSI-84-11 (1984), 389.

AN IMPROVED DEFINITION OF THE LONGITUDINAL ACCEPTANCE FOR HIGH INTENSITY BUNCHED BEAMS

M. Pabst, K. Bongardt and U. Funk
Kernforschungsanlage Jülich, D-5170 Jülich
Federal Republic of Germany

ABSTRACT

For designing high intensity linear accelerators all known effects for emittance increase should be avoided, like mismatch, beam instabilities and reorganization of the particle distribution due to the field energy. In order to have a reasonable accelerating gradient and to avoid instabilities, in many cases the longitudinal acceptance is too small at the beginning. Therefore in high intensity bunched beams particles at the end of the bunch will see strong nonlinear field components. This nonlinearity causes emittance growth but no particle losses. In this paper a model is described by which an improved calculation of the longitudinal acceptance can be given.

INTRODUCTION

A high current and a high duty cycle in proton linear accelerators means that beam losses can lead to severe problems of activation. Calculations of beam dynamics are therefore necessary in order to determine the beam parameters that will minimize these losses. Having no theory for predicting particle losses the problem is approached by designing the linac for minimum emittance growth. The emittance growth considered here results from mismatch, resonances and beam instabilities due to high space charge. In practice it turns out that for high intensity proton linacs with an input energy around 2 MeV the mechanisms which cause transverse emittance growth can be avoided by choosing appropriate parameters. But due to the high space charge the longitudinal acceptance can be much smaller than the emittance. Another design where the longitudinal acceptance is larger than the emittance corresponds to parameters which result in transverse emittance growth. Because particle losses are much more severe transversely the first design concept of choosing the parameters is the more reasonable one.

Multiparticle calculations show that a too small longitudinal acceptance will result in longitudinal emittance increase but not necessarily lead to particle losses.[1] This paper is an extended version of a previous one,[2] where a model is presented which explains that even for a longitudinal emittance/acceptance ratio of the order 10 or more no longitudinal particle loss was observed.

This effect has been studied in much detail in connection with SNQ proton Alvarez accelerator.[1,3] The main parameters of the machine are listed in Table 1.

Table 1. Parameters of the SNQ proton Alvarez linac.

Input energy	MeV	2
Output energy	MeV	102
Peak current	mA	200
Average current	mA	5
Pulse duration	μs	250
Repetition cycle	Hz	100
RF frequency	MHz	201.25
Accelerating gradient	MV/m	~ 1
Synchronous phase	deg	-32

NUMERICAL RESULTS

The basis for the new model are multiparticle calculations carried through with the program MAPRO for 2000 macroparticles. Space charge is included in all three dimensions assuming no symmetry. The space charge forces acting on a macroparticle are calculated by summing up the Coulomb forces to all other macroparticles. The beam parameters at input used for the multiparticle calculations are listed in Table 2.

Table 2. Beam parameters used for multiparticle calculations at input of the SNQ Alvarez accelerator.

Normalized transverse rms emittance	π mm mrad	1.4
Normalized longitudinal rms emittance	π-deg-MeV	0.37
Transverse tune σ_t	deg	44
Transverse tune depression σ_t/σ_{t0}		0.77
Longitudinal tune σ_L	deg	9.5
Longitudinal tune depression σ_L/σ_{LO}		0.54
Longitudinal acceptance for 200 mA	π-deg-MeV	0.19
Longitudinal acceptance for 0 mA	π-deg-MeV	3.8

The input emittances are rather large. It has been taken care of the growth rates of the injection section in front of the SNQ Alvarez linac.[3] The transverse tune σ_t has been chosen constant along the machine and equal to 44 deg to have a reasonable average beam diameter of 1.1 cm. The transverse and the longitudinal tune depression are rather small and therefore no severe space charge instabilities are expected. But the longitudinal tune depression is large enough to give a 8 times smaller acceptance than the total emittance of 1.5 π-deg-MeV. The total emittance is about 2.5 times smaller than the zero current acceptance.

In Figure 1 for the SNQ Alvarez linac the rms and 95 percent emittance growth is shown. The calculations are done with a waterbag distribution in the transverse phase space and a Gaussian and also a waterbag distribution in the longitudinal phase space. Both distributions have the same rms parameters at input. Almost "no" transverse

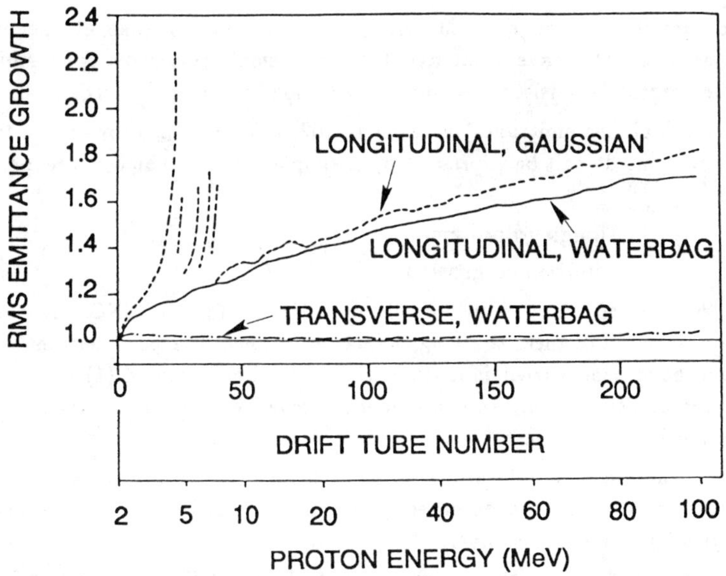

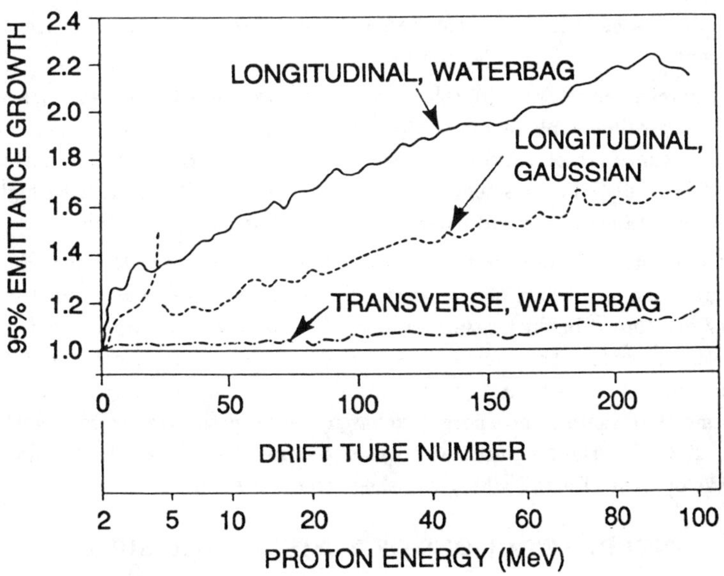

Figure 1. Rms and 95 percent emittance growth in the SNQ Alvarez linac. Input parameters are given in Table 2.

rms emittance growth and no particle loss is observed showing that all mechanisms causing emittance growth have been avoided. The small transverse rms emittance growth of some percent is consistent with the prediction by the field energy model.[4]

The longitudinal rms emittance increase at 102 MeV depends only on the longitudinal tune depression. It can be expressed by a simple formula which is precise within 10 percent:

$$\begin{matrix}\text{longitudinal rms} \\ \text{emittance growth}\end{matrix} \simeq (\sigma_{LO} / \sigma_L)^{0.9} \tag{1}$$

This formula is a fit to many multiparticle calculations for the SNQ Alvarez linac where the transverse emittance, the longitudinal emittance, the peak current and the transverse tune have been varied independently.[1] The limitation of (1) is to have no resonances or instabilities which can cause energy exchange between the transverse and longitudinal direction of motion.

Comparing the longitudinal emittance growth in Figure 1 for the Gaussian and waterbag input distributions the observed rms emittance growth is rather the same. The total longitudinal emittances differ.

The total longitudinal emittance for the waterbag distribution is 1.5 π-deg-MeV which is less than the acceptance without current. The total longitudinal emittance for the Gaussian distribution is 5.2 π-deg-MeV which is even larger than the acceptance without current. The latter case causes a particle loss of about 0.5 percent at the beginning of the linac shown in Figure 1. The model presented later on refers to the more important cases where the emittance is in between the acceptance with and without current.

The 95 percent emittance growth shows much less increase in case of a Gaussian distribution. A possible explanation of this effect is the fact that the Gaussian distribution has a four times larger total emittance at input compared to the waterbag distribution. So if both distributions tend to become asymptotically equal with higher energies the total emittance of the waterbag distribution has to increase faster.

To get more detailed information on the internal structure of the particle distribution the projections of all particles onto the z-axis and the x-axis as a function of drift tube number are shown in Figure 2a, b and c. The distributions are plotted in units of the rms width. Therefore the effects of quadrupole focusing, acceleration and emittance change are separated away. All projections do not change shape significantly. This again confirms that there is no energy exchange between the transverse and longitudinal motion. Studying the longitudinal projections one observes at the beginning of the linac the development of a tail which vanishes after some cells.

INTERPRETATIONS OF NUMERICAL RESULTS

Traditionally the longitudinal particle motion on the z-axis is described by a sum of the nonlinear rf-potential and a quadratic space charge potential:

$$V_{in}(\Delta\phi) = qE_0T\,\frac{\beta_s\lambda}{2\pi}\left\{\sin(\Delta\phi+\phi_s) - \Delta\phi\cos\phi_s - \sin\phi_s + \frac{1}{2}\mu_L\sin\phi_s\Delta\phi^2\right\} \tag{2a}$$

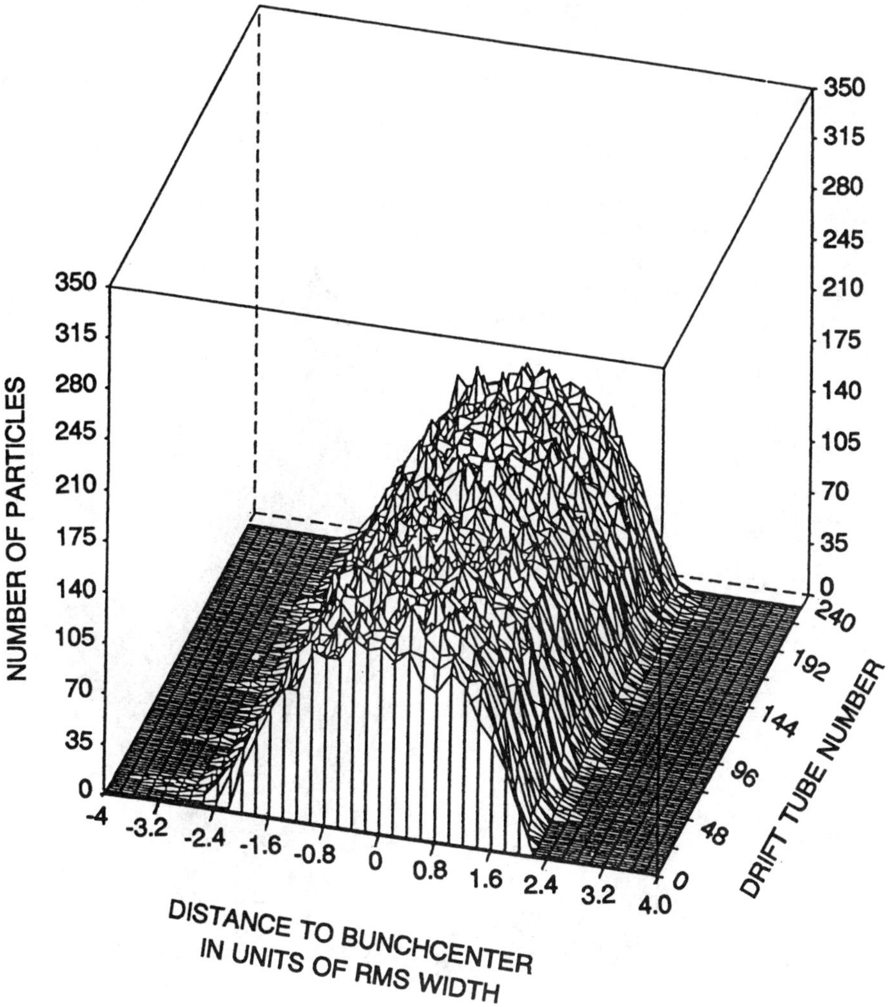

Figure 2a. Projection of the 6d phase space onto the z-axis as a function of the drift tube number.
Input distribution: transverse waterbag, longitudinal waterbag

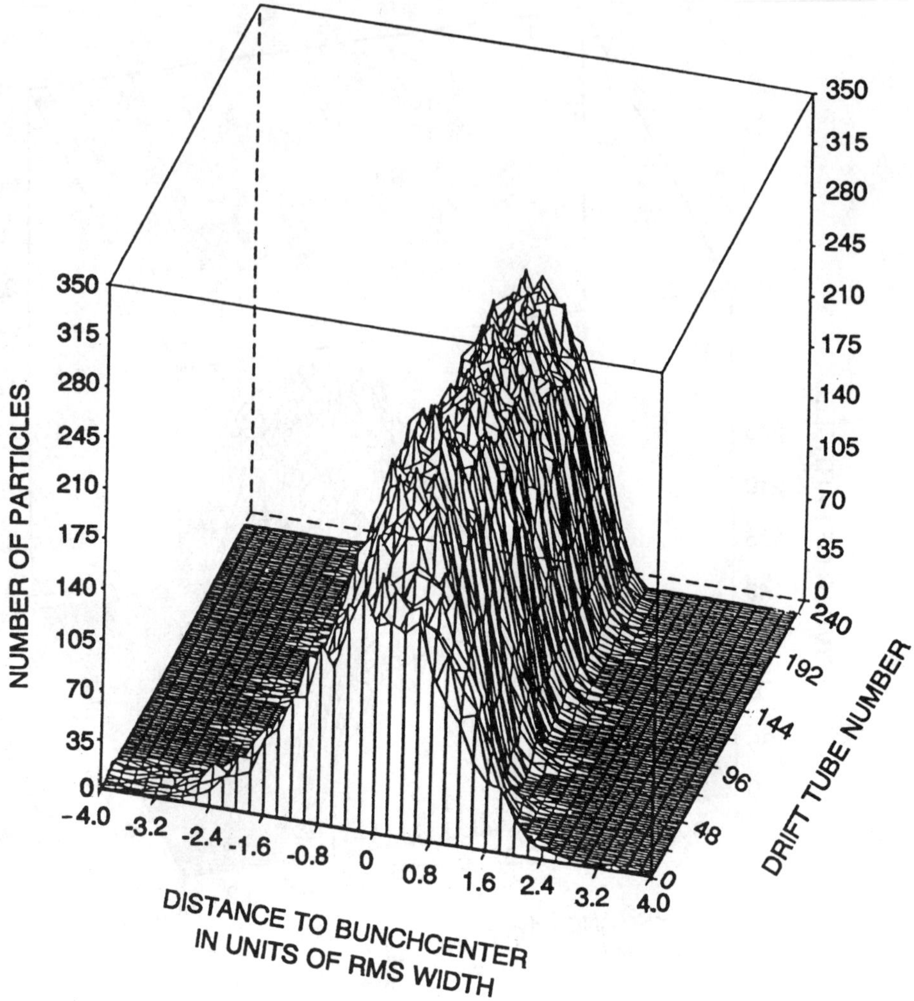

Figure 2b. Projection of the 6d phase space onto the z-axis as a function of the drift tube number

Input distribution: transverse waterbag, longitudinal Gaussian

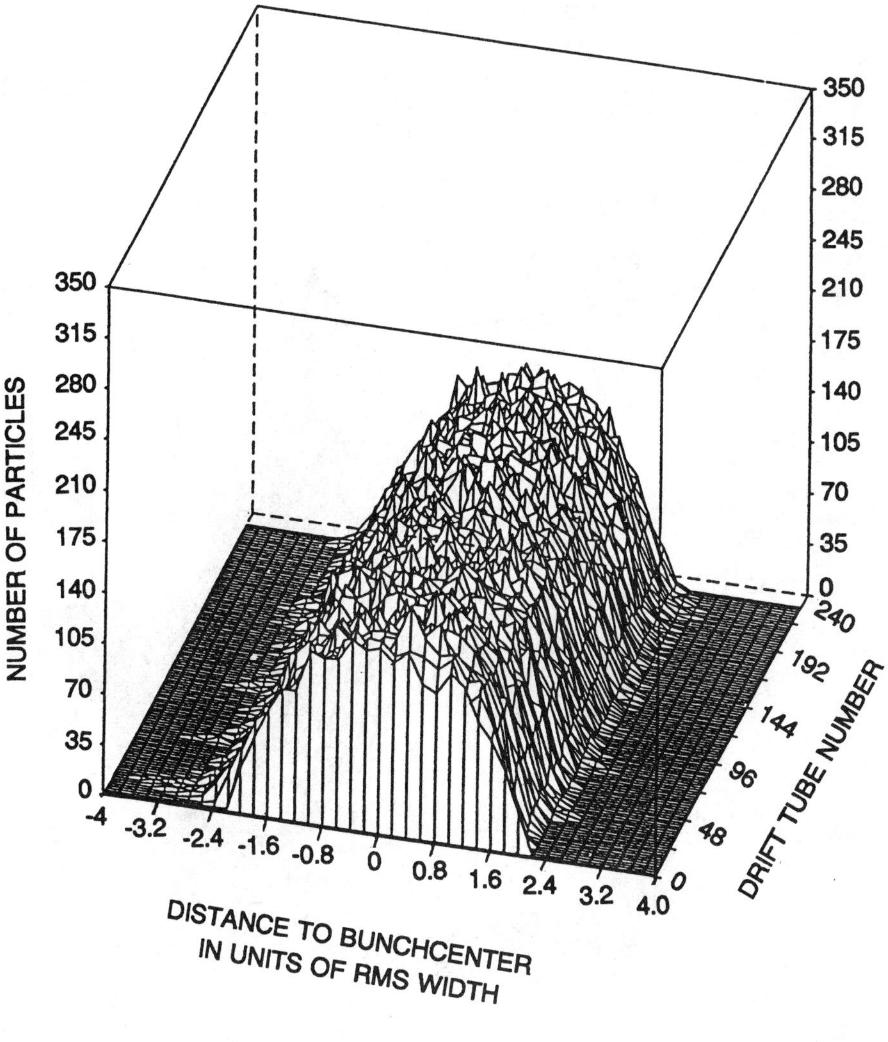

Figure 2a. Projection of the 6d phase space onto the z-axis as a function of the drift tube number.

Input distribution: transverse waterbag, longitudinal waterbag

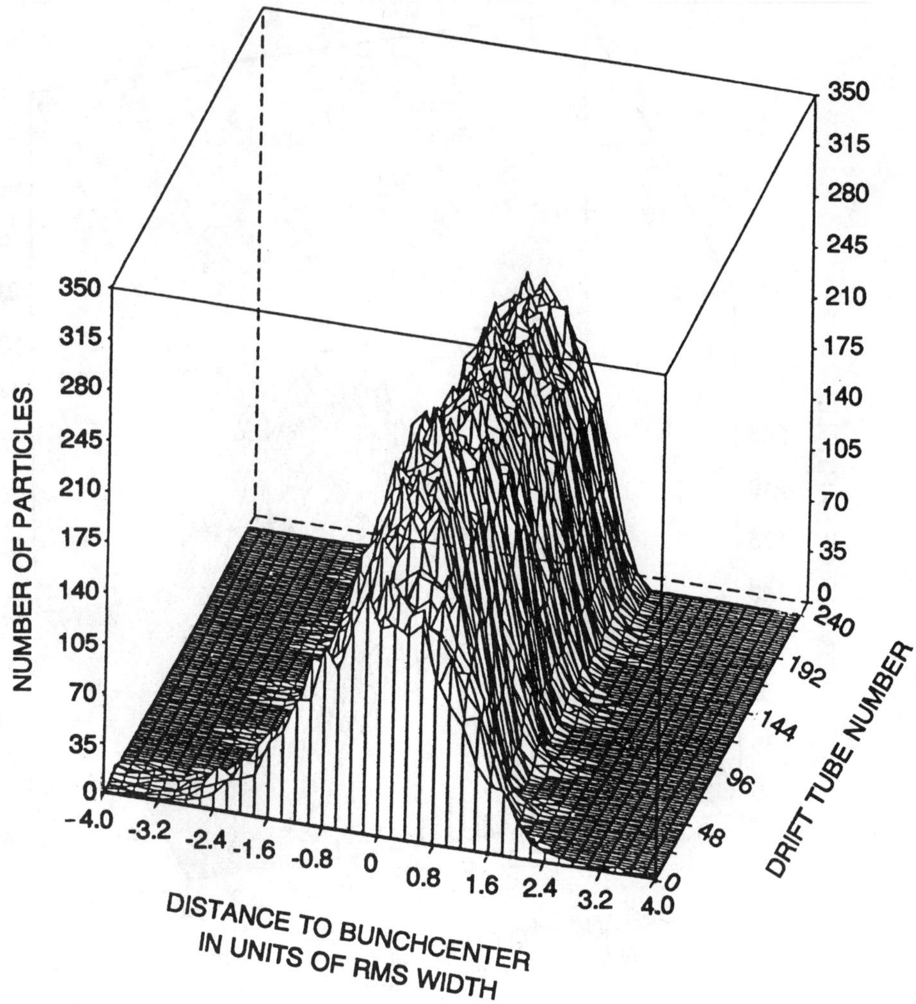

Figure 2b. Projection of the 6d phase space onto the z-axis as a function of the drift tube number

Input distribution: transverse waterbag, longitudinal Gaussian

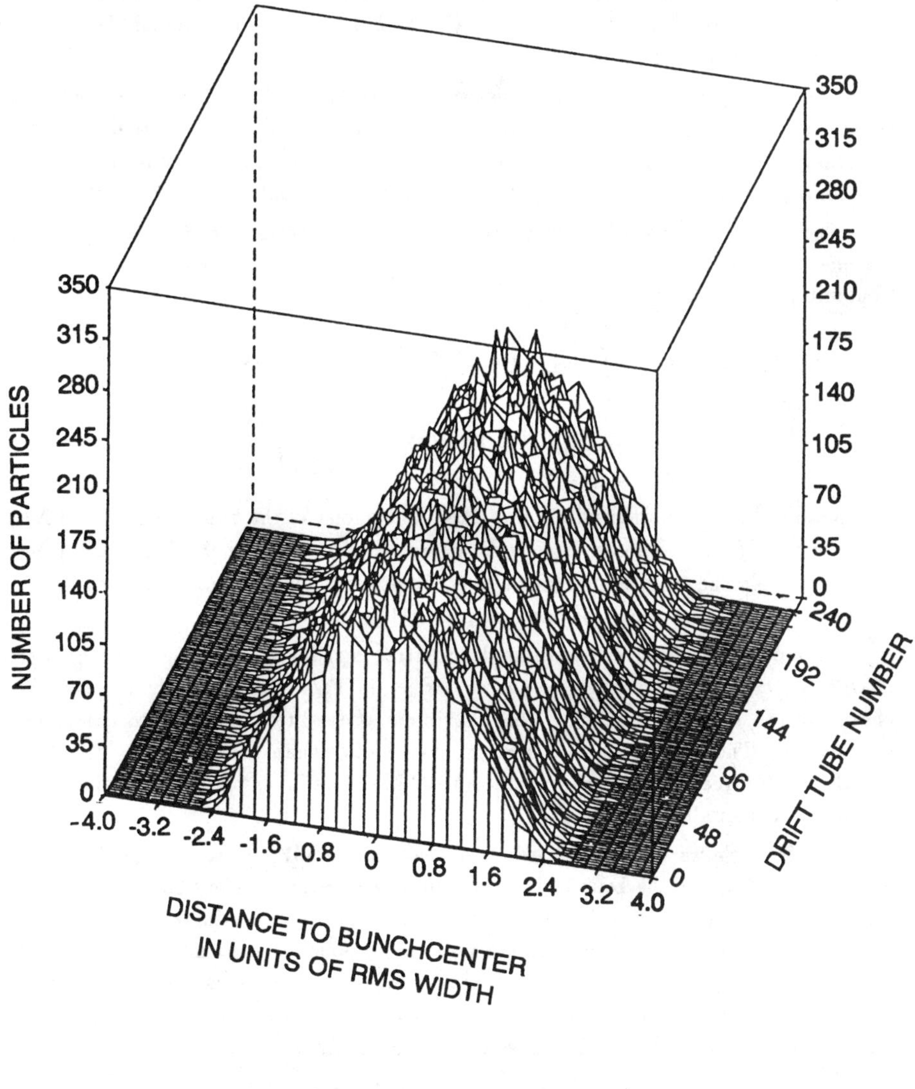

Figure 2c. Projection of the 6d phase space onto the x-axis as a function of the drift tube number

Input distribution: transverse waterbag, longitudinal waterbag

Here $\Delta\phi = \phi - \phi_s$ is the phase difference of the particle with phase ϕ to the synchronous particle with phase ϕ_s; q is the proton charge, E_0 the accelerating field, T the transit time factor, λ the rf wavelength, β_s and γ_s are the relativisitic factors. The potential is scaled to zero for $\Delta\phi = 0$.

The assumption of presenting the beam by an ellipsoid uniformly filled with space charge is used only for calculating the space charge factor $\mu_L = 1 - (\sigma_L/\sigma_{Lo})^2$. For describing the single particle motion linear space charge forces are assumed everywhere. The equation (2a) is commonly used to calculate the longitudinal acceptance[1,5,6].

The model presented here is an extension of the described model. It assumes that the effective single particle potential is equal to (2a) inside the bunch but equal to the Coulomb potential outside the bunch:

$$V_{out}(\Delta\phi) = qE_0T\,\frac{\beta_s\lambda}{2\pi}\left\{\sin(\Delta\phi+\phi_s) - \Delta\phi\cos\phi_s - \sin\phi_s + \mu_L\sin\phi_s\left[\frac{3}{2}\,|\Delta\phi_m|^2 - \frac{|\Delta\phi_m|^3}{|\Delta\phi|}\right]\right\} \tag{2b}$$

The potential and the longitudinal force are continuous at the bunch boundary $|\Delta\phi_m|$. For estimating the Coulomb force it is assumed to have a bunch of length $2|\Delta\phi_m|$. As an example the longitudinal single particle potential inside and outside the bunch is plotted in Figure 3 at input (3a) and output (3b) of the linac.

At input the effective longitudinal potential is reduced significantly by high space charge ($\mu_L \sim 0.7$). The inner potential barrier in Figure 3 results from linear space charge forces inside the bunch. This maximum is traditionally used to calculate the acceptance. The outer potential barrier, even higher than the inner one, results from the Coulomb forces outside the bunch. At 102 MeV the inner potential barrier has vanished due to a smaller bunch length.

Figure 4 shows the corresponding phase space $(\Delta W, \phi)$ at input. Three curves are plotted. No. 1 is the boundary of the initial waterbag distribution. Its emittance is 1.5 π-deg-MeV. No. 2 is the trajectory passing the inner maximum of the potential. The inner maximum corresponds to the inner hyperbolic fix point in Figure 4. Its phase is $\phi_1 = -\phi_s\,(1-\mu_L)$. The shaded area is the acceptance calculated traditionally which is much smaller than the emittance. No. 3 is the trajectory passing the outer maximum of the potential. The phase of this maximum is evaluated to be

$$\phi_2 = -\phi_s + \Delta\phi_{corr} \tag{3}$$

with

$$\Delta\phi_{corr} \simeq -\mu_L\,\frac{|\Delta\phi_m|^3}{4\phi_s^2}\cdot\frac{1}{1+\mu_L\,|\Delta\phi_m|^3/(4\phi_s^3)} \tag{4}$$

With $|\Delta\phi_m| = 28$ deg, $\mu_L = 0.7$ and $\phi_s = -32$ deg this gives $\Delta\phi_{corr} = -4.5$ deg and $\phi_2 = 27.5$ deg. The corrésponding maximum energy spread can be easily evaluated by

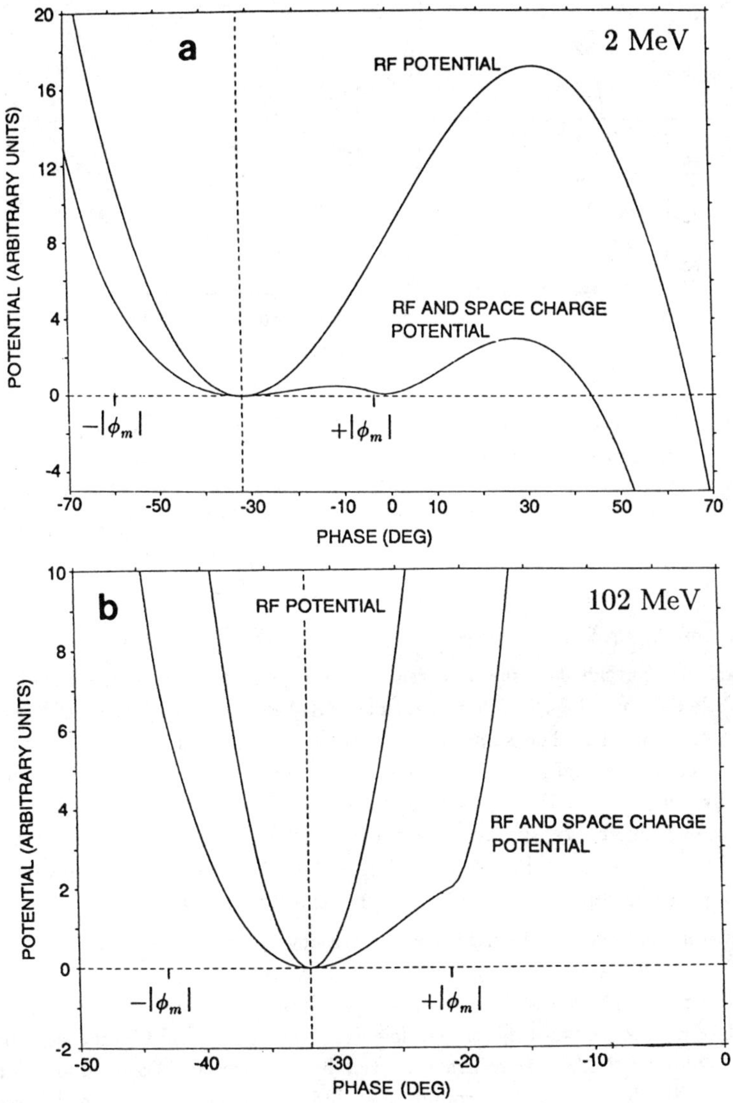

Figure 3. Effective longitudinal single particle potential inside and outside the bunch at 2 MeV and 102 MeV. The bunch length is $2\,|\Delta\phi_m|$.

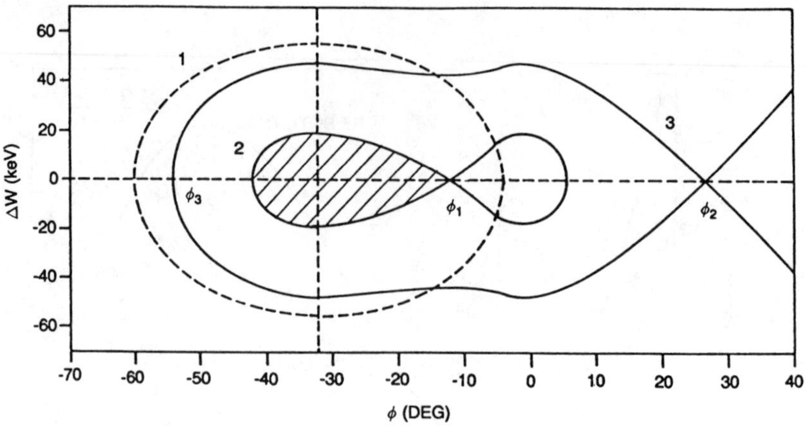

Figure 4. Longitudinal phase space at 2 MeV. The corresponding potential is shown in Figure 3a.

$$\Delta W_{\max} = 2mc^2 \, \beta_s^2 \, \gamma_s^3 \, V_{out} \, (\Delta\phi = -2\phi_s + \Delta\phi_{corr}) \quad . \tag{5}$$

Here mc^2 is the proton rest mass.

Particles outside the shaded area will move to the end of the bunch and form a tail with less space charge. These particles will be reflected by the outer potential barrier and "stabilized." The same happens to particles outside curve No. 3. Until these particles reach the end of the bunch the maximum energy spread has increased fast enough by acceleration to accept these particles. Therefore no particle loss is expected if the maximum energy spread is less than $\pm\Delta W_{\max}$ and if the half bunch length $|\Delta\phi_m|$ is smaller than $|\phi_3 - \phi_s|$. The value of ϕ_3 can be calculated numerically from the potential function given by Eq. (2a) and (2b) (see also Figure 4).

To confirm this model single particle trajectories and phase space distributions are plotted. Figure 5 shows the distribution in the longitudinal phase space at 2, 2.5 and 3 MeV. The picture at 2 MeV corresponds to Figure 3a and 4. Between 2 and 2.5 MeV a tail has evolved consisting of particles moving around the inner potential barrier. Between 2.5 and 3 MeV they are reflected by the second potential barrier. The tail particles will then oscillate around the bunch center. This can be seen also on Figure 2a and 2b.

A trajectory of a particle inside the bunch is shown in Figure 6a. It stays on the left side of the inner potential barrier and makes about 2.5 synchrotron oscillations between 2 and 102 MeV consistent with the longitudinal tune.

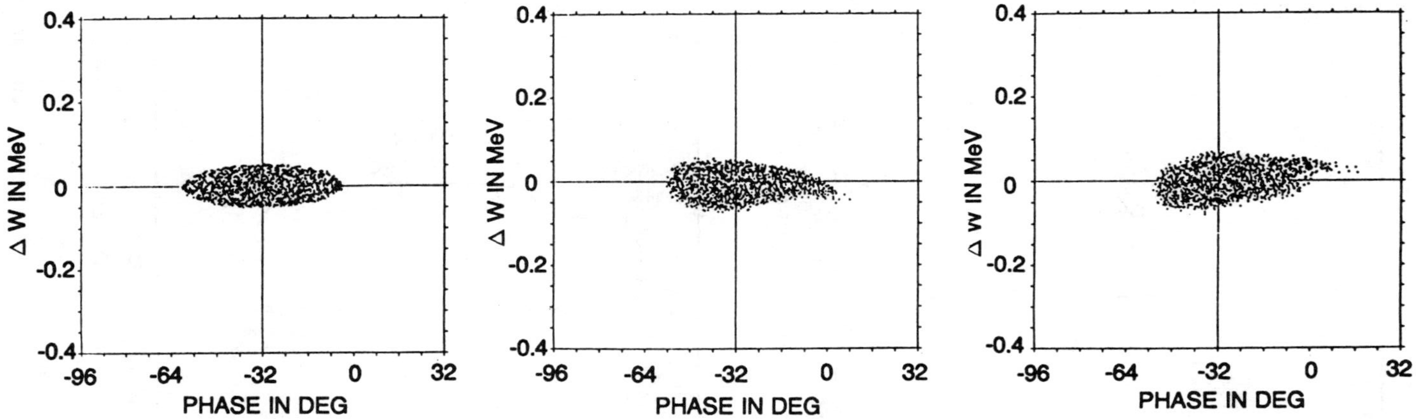

Figure 5. Projections of the particles onto the $(\phi, \Delta W)$—phase space at 2, 2.5 and 3 MeV.

Input distribution: transverse waterbag, longitudinal waterbag

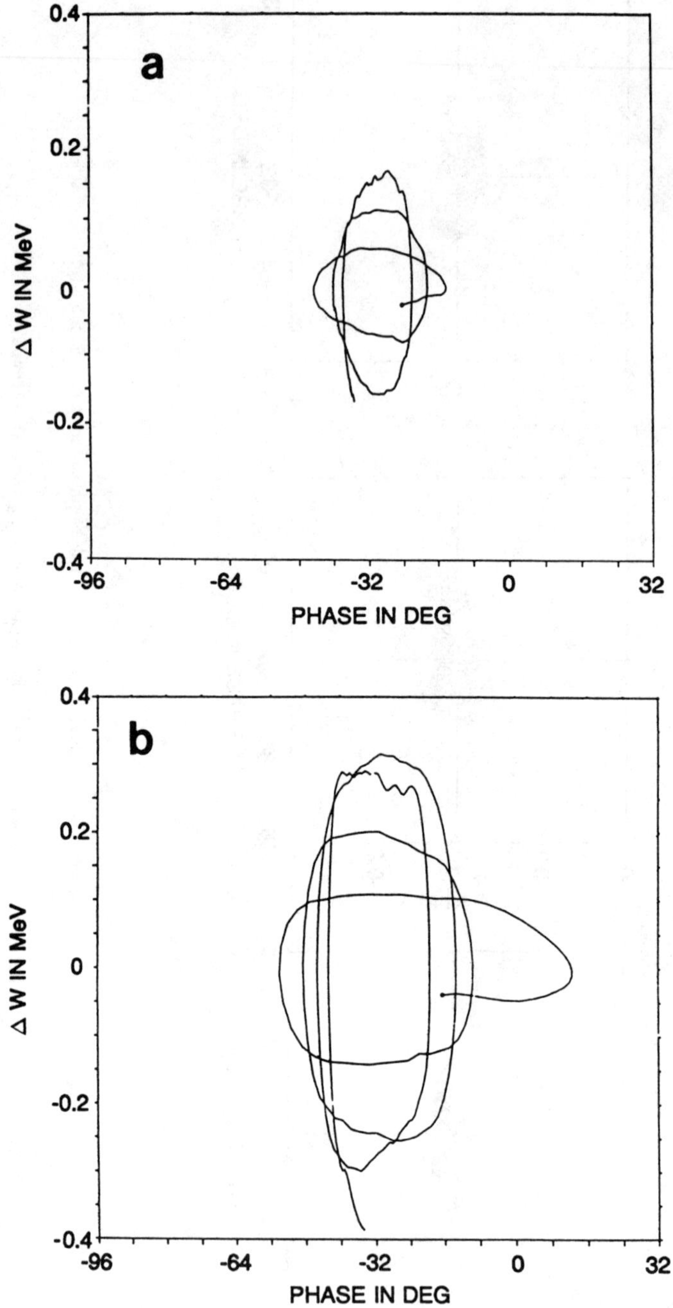

Figure 6. Single particle trajectories in the longitudinal phase space between 2 and 102 MeV.

A second particle is shown in Figure 6b which starts at the right side of the inner potential barrier and is then reflected and stabilized by the second potential barrier. When it is stabilized it sees much less space charge and therefore it oscillates faster in the phase space. It makes about 3.5 synchrotron oscillations.

The presented model is only valid if not too much particles will leave the bunch. Otherwise the space charge potential itself changes too much with the leaving particles. This model explains that for high space charge a second separatrix is outside the bunch. The outer maximum can be used to calculate a more realistic longitudinal acceptance.

CONCLUSIONS

High current linacs can be designed with almost no transverse emittance increase, no particle losses and stable particle distributions. The longitudinal emittance will increase due to high space charge but the particles are prevented from getting lost by a second potential barrier. This result is caused by the nonlinear rf potential and the nonlinear space charge potential outside the bunch. These considerations lead to a new calculation of the longitudinal acceptance.

ACKNOWLEDGEMENTS

The authors wish to express many thanks to H.G. Hereward for numerous stimulating discussions and for his excellent help in understanding the physical properties of many particle systems.

REFERENCES

1. M. Pabst, "A Design Concept for the SNQ Alvarez Linear Accelerator for Low Emittance Growth," 1984 Linac Conf., Seeheim, GSI-84-11, p. 383.
2. M. Pabst, K. Bongardt, U. Funk, "Nonlinear Field Effects and Longitudinal Emittance Growth of High Intensity Bunched Beams," to be published in Particle Accelerator Conference, Vancouver, 1985.
3. C. Zettler, "The Linear Accelerator and Pulse Compressor of the SNQ-Project," 1984 Linac Conf., Seeheim, GSI-84-11, p. 480.
4. J. Struckmeier, J. Klabunde and M. Reiser, "On the Stability and Emittance growth of Different Particle Phase-Space Distributions in a Long Magnetic Quadrupole Channel," Part. Acc. **15**, (1984), p. 47-65.
5. K. Mittag, "On Parameter Optimization for a Linear Accelerator," Kernforschungszentrum Karlsruhe, Report: KfK-2555, 1978.
6. R. Gluckstern, "Space charge effects," in Linear Accelerators, ed. P.M. Lapostolle and A.L. Septier, North-Holland (1970).

PRELIMINARY DESIGN AND BEAM-DYNAMICS STUDY OF A FUNNELING LINE HAVING LOW EMITTANCE GROWTH

F. W. Guy, AT-1, MS H817
Los Alamos National Laboratory, Los Alamos, NM 87545

ABSTRACT

A theoretical design study has resulted in a conceptual funneling-line design that has a transverse emittance growth limited to only 15% based on beam-dynamics calculations. Two 2-MeV, 100-mA proton beams are funneled from a two-channel, 212.5-MHz radio-frequency quadrupole (RFQ) to a single beam suitable for injection into a 425-MHz linac. The design uses permanent-magnet quadrupoles, dipoles, and combined-function elements. The low emittance growth is obtained by arranging the focusing strength, the periodic structure, and the bending elements so as to minimize abrupt changes in the beam environment with consequent charge redistribution and space-charge-caused emittance growth.

INTRODUCTION

Funneling lines in which two bunched beams of current I and bunch separation $\beta\lambda$ are merged into a single beam of current 2I and bunch separation $\beta\lambda/2$ may be advantageous for a number of different applications. For some of these applications, emittance growth must be kept to a minimum. Insights gained from a previous funneling-line design study,[1] recent advances in beam-dynamics theory,[2] and RFQ design experience at Los Alamos have provided a better understanding of space-charge-induced emittance growth and have shown the necessary ingredients and limitations to obtain low-emittance-growth funneling- and transport-line designs. This paper presents such a design and gives the results of a beam-dynamics study. Although the design is not optimized, the calculational results indicate that emittance growth can be controlled.

DESIGN CONSIDERATIONS

Bongardt[1] has reported the results of a funneling study completed for the German Spallation Neutron Source. Bongardt's line was designed to funnel 100-mA beams of 2-MeV protons from two 100-MHz RFQs into a 200-MHz linac. The average transverse emittance growth in the entire line, including kickers and diagnostics, was $\epsilon_f/\epsilon_i = 2.4$. Bongardt concluded that much of this growth was due to the change in particle distribution as the beam adapted itself to the funneling line. Wangler[2] showed how this occurs; the charge distribution changes rapidly if space-charge forces are not in equilibrium with external focusing forces, and the residual field energy

---

*Work was supported by the US Dept. of Energy.

for nonuniform beams (the nonlinear field energy) can be transformed to transverse energy of the particles, causing emittance increase.

If the focusing strength or the periodicity (as seen by a beam traversing the various elements of a funneling or transport line) varies so that the average radius of the beam changes, the beam charge distribution also will change as the space-charge forces seek a new equilibrium with the focusing forces. This change in charge distribution results in emittance increase. Thus, we see that if the beam environment (the average focusing and bending strengths and periodicity in the line) could be kept constant, then space-charge-induced emittance growth could be minimized. We assume[3] that changes in the beam environment, if unavoidable, should be as gradual as possible to try to approach an adiabatic redistribution of charge.

Other causes of emittance growth are energy dispersion in bending elements, spatial nonlinearities of external magnetic and electric fields, and the sine-wave time dependence of bunching and deflecting electric fields. The effect of most of these can be reduced by minimizing beam size in all three dimensions. A small beam radius reduces the effect of nonlinear fields in quadrupoles, bending dipoles, and rebunchers. A short bunch keeps particles in the linear portion of the sine-wave rebuncher field and reduces dispersion caused by the deflecting field in rf deflectors.

In the design presented in this paper, the goal was to keep the focusing strength and periodicity of the funneling line the same as the preceding RFQs; this goal was met only partially. The line was designed to keep beam dimensions as small as possible. Within the bending elements, bunch charge redistribution associated with dispersion was reduced (1) by using a two-channel RFQ with output beams separated by 2.4 cm to substantially reduce bending from that necessary for two completely separate RFQs and (2) by distributing bending throughout the line. No space allocation was made for diagnostics; space could be made available, but it would be necessary to maintain focusing strength and periodicity to avoid excessive emittance growth.

LINE ELEMENTS

A schematic drawing of the funneling line studied is shown in Fig. 1. Parameters of the line elements are given in Table I. The line is 21.5 cm long and consists of two periods. Input beam parameters were provided by an unpublished RFQ design study by Wangler and Mills.* Their design would operate at 212.5 MHz and would produce a 100-mA beam of 2-MeV protons with an average transverse normalized rms emittance of 0.022 π•cm•mrad. In this study, I assumed that the two RFQ output beams would be identical except that the bunches would be 180° out of phase, which would be accomplished by proper phasing of the vane or rod modulations in each RFQ beamline.

*This information supplied by T. P. Wangler and R. S. Mills, AT-1, Los Alamos National Laboratory.

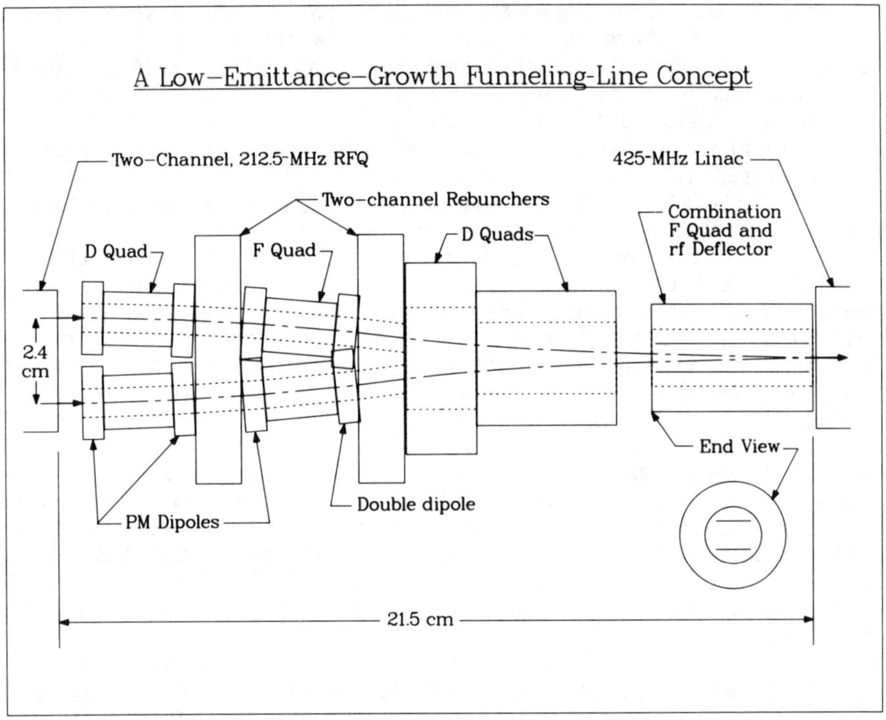

Fig. 1. Drawing of funneling line.

Table I. Beamline elements

1. Drift, 0.7 cm long
2. Dipole, 0.6 cm long, -1.69° bend, 10 000 G
   (All dipoles and quadrupoles are permanent magnets)
3. Defocusing quadrupole, 2.0 cm long, 20 000-G/cm field gradient
4. Dipole, 0.6 cm long, -1.69° bend, 10 000 G
5. Rebuncher, 1.4 cm long, 0.12-MV peak voltage, 425 MHz
6. Dipole, 0.6 cm long, -1.69° bend, 10 000 G
7. Focusing quadrupole, 2.0 cm long, 24 000-G/cm field gradient
8. Dipole, 0.6 cm long, -1.69° bend, 10 000 G; a "double dipole"
9. Rebuncher, 1.4 cm long, 0.04-MV peak voltage, 425 MHz
10. Defocusing quadrupole, 2.0 cm long, 7 000-G/cm field gradient
11. Defocusing quadrupole, 4.0 cm long, 10 000-G/cm field gradient
    (The preceding two quadrupoles bend the beam a total of +3.9°)
12. Drift, 1.0 cm long
13. Combination element, 4.6 cm long:
    a. Focusing quadrupole, 12 000-G/cm field gradient
    b. Rf deflector, 6.5-MV/M peak voltage, 212.5 MHz; +2.86° bend

Dimensions of the magnetic elements and of the beamline portions of the rf elements are shown to relative scale in Fig. 1. Inductive portions of the rf elements are not shown because their configurations have not been determined. All magnetic elements in the line are permanent magnets; dimensions of these elements are calculated from the formulae given by Halbach[4] subject to the conditions that (a) the material remanant field B_r = 12 kG, (b) the maximum pole tip field is 10 kG, and (c) the magnets are each composed of eight segments of permanent-magnet material.

The first period of the line is 1 βλ (9.2 cm) long. It consists of a defocusing quadrupole, a focusing quadrupole, four dipoles distributed along the line to bend the beam towards the final axis, and two 425-MHz, two-channel rebunchers to compress the bunch longitudinally. It is necessary to operate the rebunchers at 425 MHz because bunches transit the rf gaps at this frequency, passing alternately through the two channels. The bunch length must be reduced about a factor of 2 along the length of the funneling line to match the beam to the following linac because the frequency of the linac is twice the frequency of the RFQs. The period ends at the center of the second rebuncher. The element axes, including the rebuncher channel axes, are centered on the individual beam paths. The first quadrupole of the funneling line is defocusing in the x-plane (the plane of the bends) and focusing in the y-plane to match the RFQ output. The beam is bent inward 1.69° at each dipole. There is not enough separation between the beams at the last dipole to permit implementation of conventional design procedures. Therefore, a combined dipole magnet (the "double dipole" in Fig. 1) similar to the "septumless septum magnet" suggested by Halbach* is used. The double dipole is shaped like a figure eight without a center bar rather than like the circle of the normal permanent-magnet dipole. The separation of the bending and focusing functions into distinct components is convenient for calculational purposes and may be the best way to construct the actual magnets. The bending function also could be done by offsetting the quadrupoles from the beam axes or by including a dipole component in the quadrupole when the magnetic segments are specified, manufactured, and assembled. Apertures along the first period are 0.4 cm, as shown by dotted lines in Fig. 1.

The second-period component axes are centered on the final output beam axis. The first element in this period is a combination of permanent-magnet defocusing quadrupoles that provide as much focusing strength as possible while simultaneously deflecting the beam back toward the direction of the final axis. The beam exits this element with proper angular direction (x') and x-offset from the axis to intersect the axis in the center of the next element, which is a combined permanent-magnet focusing quadrupole and rf electric-field deflector. In the combined element, the beam is

---

*This information was supplied by K. Halbach, University of California, Lawrence Berkeley Laboratory, Berkeley, CA 94720.

focused in the x-direction and deflected along the output z-axis. The deflector operates at 212.5 MHz to provide the proper alternating deflection angles for bunches coming alternately from the two input lines. Apertures in the second period vary to accommodate the changing offset of the beam from the beamline. Separation between the rf deflector plates is 0.8 cm; therefore, the final radial aperture is 0.4 cm. Ideally, this second period should be 9.2 cm long to maintain constant periodicity. However, the length of the rf deflector along the beamline is $\beta\lambda/2$ (4.6 cm) to maximize available deflection for a given peak electric field and to minimize the effect of fringe fields at the ends of the element. The defocusing quadrupole combination plus the following drift space is 7 cm long to obtain the proper beam trajectory. Half the length of the second rebuncher, 0.7 cm, is included in the second period. Therefore, the total length of the second period is 12.3 cm. The average focusing strength is less than in the first period, although focusing should really be stronger than in the first period to balance the higher space-charge forces resulting from longitudinal compression of the bunch. Better techniques for calculating the beam trajectory and designing the magnets may allow a reduction of the length of the period and an increase in focusing strength. It is important not to disturb the charge distribution of the beam; possibly the distribution can be kept constant by proper combinations of period length and focusing strength.

There are several elements in this design that have not been used before. A two-channel RFQ has not yet been demonstrated, but there seems to be no obvious reason as to why it should not perform as well as a conventional RFQ. The permanent-magnet dipoles and first-period quadrupoles are very small but should present no unexpected problems because the field requirements are not extreme. Higher order multipoles in the permanent magnets will probably be a few per cent at the radius of the aperture, and this should not have a large effect on emittance growth. The two-channel rebunchers may present a design challenge in that they must fit into a very small length of beamline and must accommodate two angled beams with the minimum possible field nonlinearity. The amount of field nonlinearity in the rebunchers has not been investigated.

The rf deflector surrounded by a focusing quadrupole is probably the highest risk component. The combination of deflecting and focusing functions was influenced by the desirability of a $\beta\lambda/2$ deflector and the simultaneous requirement of strong focusing. The final design of this element may not look like the conceptual sketch in Fig. 1. There have been only preliminary discussions and ideas concerning coupling the rf power, minimizing electric-field nonlinearities and edge effects, controlling sparking and multipactoring, and keeping heating to an acceptable level.

BEAM-DYNAMICS CALCULATIONS AND RESULTS

The beam-dynamics calculation was performed on the particle-tracking code PARMILA. A 3-D space-charge subroutine was used so that the off-axis beam (which is not expected to be round) could be calculated properly. No beam-beam or bunch-bunch interactions were included; these effects are expected to be small and can be easily included later, if desired. No image-charge effects were included. Also, no spatial nonlinearities were included in the magnetic or electric fields because these were unknown. Included in the calculation were the temporal nonlinearity in the rebuncher voltages and phase-angle-dependent and energy-dependent dispersion in the rf deflector. The effect of the rf deflector would actually occur over the entire length of the element; but in the calculations, the effect was applied as an angle impulse to each particle at the center of the element. This caused a small error that can be corrected by a more sophisticated combination deflector-quadrupole subroutine.

The funneling-line input beam was obtained by taking individual output particle coordinates from the RFQ calculation. The RFQ emitted 329 macroparticles and none were lost in the funneling-line calculation. In the first period, 100% of the macroparticles remained within a radius of 0.28 cm and 90% remained within 0.2 cm; in the entire line, 100% of the macroparticles remained within 70% of the aperture radius and 90% remained within 50% of the aperture radius. A run using the RFQ beam-ellipse parameters with an initially spatially uniform bunch of 500 macroparticles gave practically the same results for the beam envelope and emittance growth.

Beam envelopes and emittances are shown in Figs. 2 and 3. A line diagram is shown at the bottom of these figures. The quadrupoles are divided into short sections for calculational purposes. The transverse envelopes are fairly constant in the first period but vary in the second; the transverse emittances show little change in the first period and then increase in the second. If the envelope variations in the second period could be reduced, perhaps this increase could be controlled, as discussed previously. There is a small jump in the x-emittance at the position of the rf deflector that is correct in magnitude but should be distributed over the length of the deflector. The longitudinal or z-emittance shows jumps at the rebunchers (caused by the nonlinear longitudinal fields) and variations in the bending elements. These variations have not yet been fully analyzed but are believed to be caused by the mixing that bending elements introduce between longitudinal and transverse phase space. Z-emittances are shown in units of π•cm•mrad for direct comparison with transverse emittances.

The bunch length is reduced by less than a factor of 2 by the time the bunch enters the rf deflector. Bunching causes longitudinal emittance growth because the bunch is twice as long in phase angle in the 425-MHz rebunchers as in the 212.5-MHz RFQ. In future RFQ designs, if the bunch compression could be initiated in the last part of the RFQ, then the bunch could be shorter in the rebunchers,

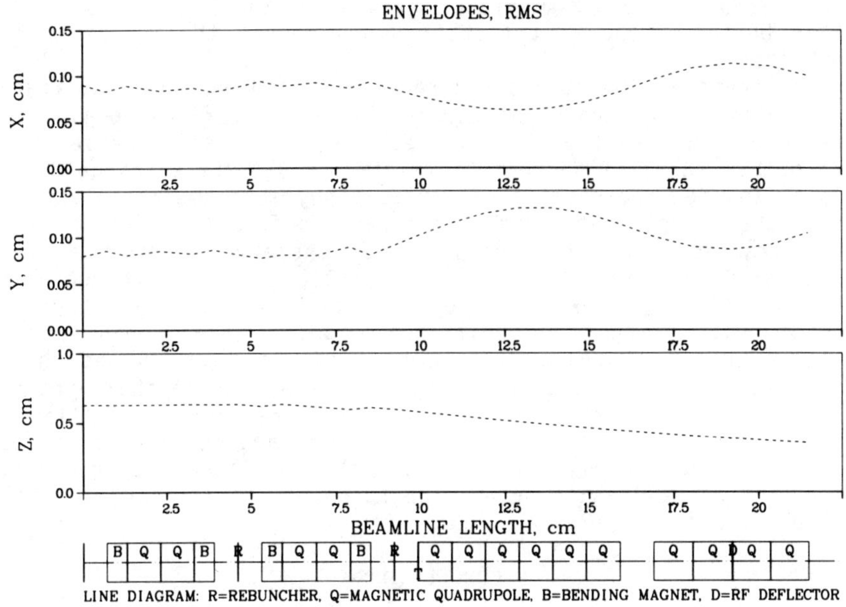

Fig. 2. Beam envelopes.

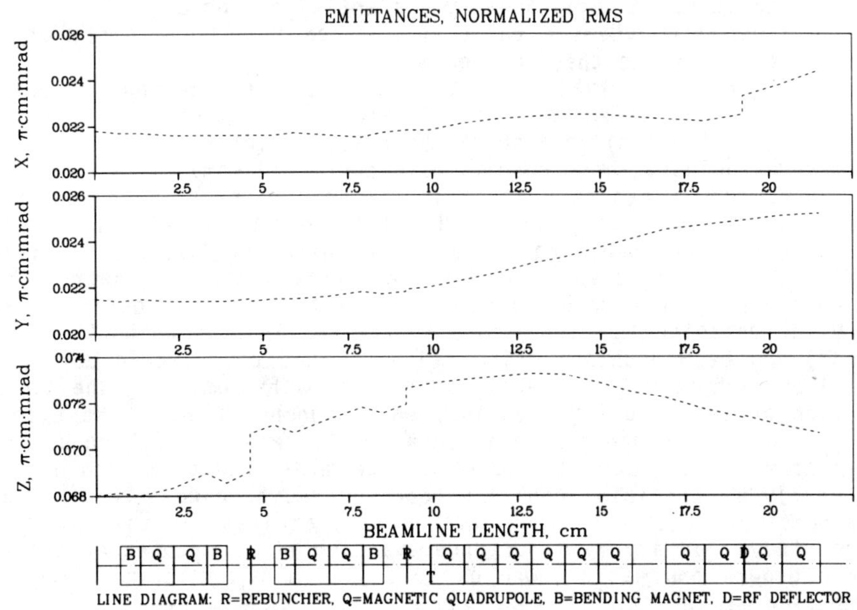

Fig. 3. Beam emittances.

thereby reducing the z-emittance growth in these elements. A shorter bunch in the rf deflector would reduce the x-emittance growth there.

The transverse emittance increase averaged over the two transverse planes in this calculation was 15%, and the final longitudinal increase was 4%. Part of this increase may have been because of the small number (329) of macroparticles in the 3-D space-charge calculation. Too few macroparticles causes unphysical small-scale lumpiness in the charge distribution that results in artificial increases in the transverse particle velocities. There is another way of looking at this effect: If two macroparticles happen to approach each other closely, they will repel each other violently, causing large velocity changes and emittance growth; whereas, the actual particles would be subjected to a more gentle repulsion as they drift into regions of higher charge density and would therefore not change velocity as much. Haber[5] has called this effect "collisional emittance growth." Using a larger number of macroparticles in both the RFQ and the funneling-line calculations will help to quantify this effect as well as to reduce the (probably fairly large) statistical uncertainty.

CONCLUSIONS

The funneling line presented in this paper gives an encouragingly low emittance growth based on beam-dynamics calculations. The study is preliminary, with the following shortcomings:

- Some of its elements may be difficult or impractical to construct exactly as they are envisioned here.
- The beam-dynamics calculations are not as complete or extensive as they should be.
- There are certain effects (such as the relation between focusing strengths, periodicity, beam radius and charge distribution) that should be better quantified.

Nevertheless, these results have confirmed some ideas about how emittance growth can be controlled. Energy dispersion in bending elements and phase-difference-induced dispersion in rf deflectors can be minimized by making the bends as small as possible. Phase-difference-induced dispersion in rf deflectors can also be reduced by compressing the bunch longitudinally (this compression is necessary for matching the bunch to a following linac in which the frequency has been doubled). The radius of the beam should be kept small by high focusing strength so that it will remain in the linear portion of the focusing, bending, and rebuncher fields. Most important of all, space-charge-induced emittance growth (which usually dominates in high-current beams) can be minimized by maintaining, as nearly as possible, constant focusing and bending strength and periodicity. The success of these ideas, at least in this simulation study, has improved our understanding of the causes of emittance growth and has allowed us to design a low-emittance-growth funneling line.

ACKNOWLEDGMENTS

The author wishes to thank K. R. Crandall and R. S. Mills for their help on PARMILA; K. Halbach, J. M. Potter, R. I. Schermer, and R. H. Stokes for helpful discussions on how the necessary hardware might be designed and constructed; and especially T. P. Wangler for his guidance and encouragement.

REFERENCES

1. K. Bongardt, Proc. 1984 Linac Conf., Gesellschaft für Schwerionenforschung, Darmstadt report GSI-84-11 (1984), 389.
2. T. P. Wangler, "Field Energy and RMS Emittance," these proceedings and Proc. 1985 Particle Accelerator Conference, Vancouver, B.C., to be published.
3. F. W. Guy and T. P. Wangler, "Funneling: An Initial Beam Dynamics Study," these proceedings.
4. K. Halbach, Nucl. Instr. and Meth. 169 (1980), 1.
5. I. Haber, "High Current Simulation Codes," these proceedings.

LIST OF PARTICIPANTS

Dr. Klaus Bongardt
KFA Jülich
SNQ-ABT, D-5170
Jülich, West Germany
49 (2461) 61-3544

Major Jerry Bueck
Air Force Weapons Laboratory/AWYW
Kirtland AFB, New Mexico 87117-6008
1(505) 844-1781.

Dr. R.J. Burke
W/B-41, Westinghouse Hanford Company
P.O. Box 1970
Richland, Washington 99352
1(509) 376-5127.

Dr. Robert Carman
Rockwell International Corporation
Rocketdyne Division
MS: 055FA12
6633 Canoga Ave.
Canoga Park, California 91303
1(818) 700-4701.

Dr. David D. Chamberlin
MS H821, Los Alamos National Laboratory
Los Alamos, New Mexico 87545
1(505) 667-3499.

Dr. Wayne Cornelius
MS H821, Los Alamos National Laboratory
Los Alamos, New Mexico 87545
1(505) 667-2966.

Dr. Kenneth R. Crandall
MS H817, Los Alamos National Laboratory
Los Alamos, New Mexico 87545
1(505) 667-1958.

Mr. N. J. Diserens
Rutherford Appleton Laboratory
Chilton, Didcot
Oxon OX11 0QX, United Kingdom
44 (235) 21900.

Dr. Thomas J. Fessenden
LBL 47-112
Lawrence Berkeley Laboratory
Berkeley, California 94720
1(415) 486-5356.

Dr. Raymond Fisher
MS/02-338, GA Technologies, Inc.
P.O. Box 85608
San Diego, California 92138
1(619) 455-4478.

Dr. George H. Gillespie
Physical Dynamics, Inc.
4350 Executive Drive, Suite 135
San Diego, California 92121
1(619) 457-3201.

Dr. Pierre Grand
Building 701, Brookhaven National Laboratory
Upton, New York 11973
1(516) 282-2038.

Dr. Frank W. Guy
MS H817, Los Alamos National Laboratory
Los Alamos, New Mexico 87545
1(505) 667-9137.

Dr. Irving Haber
Naval Research Laboratory, Code 4790
Washington, DC 20375
1(202) 767-3198.

Mr. Milton Hawie
Army Ballistic Missile Defense
Advanced Technology Center, Code ATC-E
P.O. Box 1500
Huntsville, Alabama 35807
1(205) 895-3220.

Dr. Thomas D. Hayward
The Boeing Aerospace Co., Mail Stop 3A-91
P.O. Box 3999
Seattle, Washington 98124
1(206) 773-8067.

Prof. Peter Junior
Institut fur Angewandte Physik
University of Frankfurt, Robert-Mayer-Str. 2-4
D-6000 Frankfurt am Main 1
West Germany, 49 (69) 7983485

Dr. Denis Keefe
Lawrence Berkeley Laboratory
1 Cyclotron Road
Berkeley, California 94720
1(415) 486-6376.

Dr. Roderich Keller
GSI, Gesellschaft Fur
Schwerionenforschung mbH
Planckstrassel
Postfach 110541
D-6100 Darmstadt 11
West Germany
49 (6151) 35-9321.

Dr. Charles H. Kim
Building 47, Lawrence Berkeley Laboratory
Berkeley, California 94720
1(415) 486-7218.

Dr. Jürgen Klabunde
GSI Darmstadt
Postfach 11 05 41
D-6100 Darmstadt
West Germany
49 (6151) 359-344.

Prof. Horst Klein
Institute fur Angewandte Physik
University of Frankfurt
Robert-Mayer-Str. 2-4
D-6000 Frankfurt am Main 1
West Germany
49 (69) 798-3489.

Dr. H. B. Knowles
Rockwell International Corporation
Rocketdyne Division, MS FA 38
6633 Canoga Avenue
Canoga Park, California 91303
1(818) 700-4554.

Dr. Patrick Krejcik
KFA Jülich, SNQ-ABT
D-5170 Jülich
West Germany
49 (2461) 61-3541.

Dr. Yu-Yun Kuo
Physical Dynamics, Inc.
4350 Executive Drive, Suite 135
San Diego, California 92121
1(619) 457-3201.

Dr. Edward P. Lee
Lawrence Berkeley Laboratory
1 Cyclotron Road
Berkeley, California 94720
1(415) 486-7345.

Dr. Walter P. Lysenko
MS H829, Los Alamos National Laboratory
Los Alamos, New Mexico 87545
1(505) 667-7431.

Dr. Alfred W. Maschke
TRW MS01/1020
1 Space Park
Redondo Beach, California 90278
1(213) 535-2310.

Captain Matthew G. McHarg
Air Force Weapons Laboratory/AWYW
Kirtland AFB
New Mexico 87117-6008
1(505) 846-4987.

Dr. G. E. McMichael
Atomic Energy of Canada
Chalk River Nuclear Laboratory
Chalk River, Ontario K0J 1J0 Canada
1(613) 584-3311.

Dr. Brian Murphy
Building 4500 North
Oak Ridge National Laboratory
Oak Ridge, Tennessee 37831
1(615) 574-5441

Dr. Krsto Prelec
Building 911B
Brookhaven National Laboratory
Upton, New York 11973
1(516) 282-4777.

Dr. C.R. Prior
Rutherford Appleton Laboratory
Chilton Didcot, Oxfordshire OX11 0QX
United Kingdom
44 (235) 21900 Ext. 5421.

Prof. Martin P. Reiser
Laboratory for Plasma & Fusion Energy Studies
University of Maryland
College Park, Maryland 20742
1(301) 454-3188.

Dr. Oscar R. Sander
MS H818, Los Alamos National Laboratory
Los Alamos, New Mexico 87545
1(505) 667-1012.

Dr. J. David Schneider
MS H821, Los Alamos National Laboratory
Los Alamos, New Mexico 87545
1(505) 667-5454.

Dr. S. O. Schriber
AT Division, MS H811
Los Alamos National Laboratory
Los Alamos, New Mexico 87545
1(505) 667-7634.

Dr. M. R. Shubaly
Los Alamos National Laboratory
AT Division, MS H 817
Los Alamos, New Mexico 87545
1(505) 667-9124.

Dr. Marshall M. Sluyter
U.S. Department of Energy
Office of Inertial Fusion, DP-231
Washington, DC 20545
1(301) 353-5249.

Dr. Lloyd Smith
47/112, Lawrence Berkeley Laboratory
1 Cyclotron Road
Berkeley, California 94720
1(415) 486-7206.

Dr. John Staples
Building 64, Room 224-A
Lawrence Berkeley Laboratory
University of California
Berkeley, California 94720
1(415) 486-5831.

Dr. Larry Stewart
GA Technologies
P.O. Box 85608
San Diego, California 92138
1(619) 455-4281.

Dr. W. L. Stirling
Building 9201-2, MS 2
Oak Ridge National Laboratory
P.O. Box Y
Oak Ridge, Tennessee 37831
1(615) 574-1123.

Dr. Richard H. Stokes
MS H817, Los Alamos National Laboratory
Los Alamos, New Mexico 87545
1(505) 667-9125.

Dr. R. Thomae
FOM, Institute for Atomic and Molecular Physics
P.O. Box 41883
1009DB, Amsterdam
The Netherlands
31 (20)94-6711.

Dr. Noboru Tokuda
Institute for Nuclear Study
University of Tokyo
Midori-cho 3-2-1, Tanashi-Shi
Tokyo 188, Japan
81 (424)61-4131.

Dr. E. Alan Wadlinger
MS H818, Los Alamos National Laboratory
Los Alamos, New Mexico 87545
1(505) 667-2797.

Dr. Thomas P. Wangler
MS H817, Los Alamos National Laboratory
Los Alamos, New Mexico 87545
1(505) 667-3200.

Dr. John H. Whealton
Building 9201-2, MS 2
Oak Ridge National Laboratory
P.O. Box Y
Oak Ridge, Tennessee 37831
1(615) 574-1130.

Dr. Douglas C. Wilson
DRP/FRA, MS E527
Los Alamos National Laboratory
Los Alamos, New Mexico 87545
1(505) 667-3612.

AIP Conference Proceedings

		L.C. Number	ISBN
No. 1	Feedback and Dynamic Control of Plasmas – 1970	70-141596	0-88318-100-2
No. 2	Particles and Fields – 1971 (Rochester)	71-184662	0-88318-101-0
No. 3	Thermal Expansion – 1971 (Corning)	72-76970	0-88318-102-9
No. 4	Superconductivity in *d*- and *f*-Band Metals (Rochester, 1971)	74-18879	0-88318-103-7
No. 5	Magnetism and Magnetic Materials – 1971 (2 parts) (Chicago)	59-2468	0-88318-104-5
No. 6	Particle Physics (Irvine, 1971)	72-81239	0-88318-105-3
No. 7	Exploring the History of Nuclear Physics – 1972	72-81883	0-88318-106-1
No. 8	Experimental Meson Spectroscopy –1972	72-88226	0-88318-107-X
No. 9	Cyclotrons – 1972 (Vancouver)	72-92798	0-88318-108-8
No. 10	Magnetism and Magnetic Materials – 1972	72-623469	0-88318-109-6
No. 11	Transport Phenomena – 1973 (Brown University Conference)	73-80682	0-88318-110-X
No. 12	Experiments on High Energy Particle Collisions – 1973 (Vanderbilt Conference)	73-81705	0-88318-111–8
No. 13	π-π Scattering – 1973 (Tallahassee Conference)	73-81704	0-88318-112-6
No. 14	Particles and Fields – 1973 (APS/DPF Berkeley)	73-91923	0-88318-113-4
No. 15	High Energy Collisions – 1973 (Stony Brook)	73-92324	0-88318-114-2
No. 16	Causality and Physical Theories (Wayne State University, 1973)	73-93420	0-88318-115-0
No. 17	Thermal Expansion – 1973 (Lake of the Ozarks)	73-94415	0-88318-116-9
No. 18	Magnetism and Magnetic Materials – 1973 (2 parts) (Boston)	59-2468	0-88318-117-7
No. 19	Physics and the Energy Problem – 1974 (APS Chicago)	73-94416	0-88318-118-5
No. 20	Tetrahedrally Bonded Amorphous Semiconductors (Yorktown Heights, 1974)	74-80145	0-88318-119-3
No. 21	Experimental Meson Spectroscopy – 1974 (Boston)	74-82628	0-88318-120-7
No. 22	Neutrinos – 1974 (Philadelphia)	74-82413	0-88318-121-5
No. 23	Particles and Fields – 1974 (APS/DPF Williamsburg)	74-27575	0-88318-122-3
No. 24	Magnetism and Magnetic Materials – 1974 (20th Annual Conference, San Francisco)	75-2647	0-88318-123-1

No. 25	Efficient Use of Energy (The APS Studies on the Technical Aspects of the More Efficient Use of Energy)	75-18227	0-88318-124-X
No. 26	High-Energy Physics and Nuclear Structure – 1975 (Santa Fe and Los Alamos)	75-26411	0-88318-125-8
No. 27	Topics in Statistical Mechanics and Biophysics: A Memorial to Julius L. Jackson (Wayne State University, 1975)	75-36309	0-88318-126-6
No. 28	Physics and Our World: A Symposium in Honor of Victor F. Weisskopf (M.I.T., 1974)	76-7207	0-88318-127-4
No. 29	Magnetism and Magnetic Materials – 1975 (21st Annual Conference, Philadelphia)	76-10931	0-88318-128-2
No. 30	Particle Searches and Discoveries – 1976 (Vanderbilt Conference)	76-19949	0-88318-129-0
No. 31	Structure and Excitations of Amorphous Solids (Williamsburg, VA, 1976)	76-22279	0-88318-130-4
No. 32	Materials Technology – 1976 (APS New York Meeting)	76-27967	0-88318-131-2
No. 33	Meson-Nuclear Physics – 1976 (Carnegie-Mellon Conference)	76-26811	0-88318-132-0
No. 34	Magnetism and Magnetic Materials – 1976 (Joint MMM-Intermag Conference, Pittsburgh)	76-47106	0-88318-133-9
No. 35	High Energy Physics with Polarized Beams and Targets (Argonne, 1976)	76-50181	0-88318-134-7
No. 36	Momentum Wave Functions – 1976 (Indiana University)	77-82145	0-88318-135-5
No. 37	Weak Interaction Physics – 1977 (Indiana University)	77-83344	0-88318-136-3
No. 38	Workshop on New Directions in Mossbauer Spectroscopy (Argonne, 1977)	77-90635	0-88318-137-1
No. 39	Physics Careers, Employment and Education (Penn State, 1977)	77-94053	0-88318-138-X
No. 40	Electrical Transport and Optical Properties of Inhomogeneous Media (Ohio State University, 1977)	78-54319	0-88318-139-8
No. 41	Nucleon-Nucleon Interactions – 1977 (Vancouver)	78-54249	0-88318-140-1
No. 42	Higher Energy Polarized Proton Beams (Ann Arbor, 1977)	78-55682	0-88318-141-X
No. 43	Particles and Fields – 1977 (APS/DPF, Argonne)	78-55683	0-88318-142-8
No. 44	Future Trends in Superconductive Electronics (Charlottesville, 1978)	77-9240	0-88318-143-6
No. 45	New Results in High Energy Physics – 1978 (Vanderbilt Conference)	78-67196	0-88318-144-4

No.	Title	LC No.	ISBN
No. 46	Topics in Nonlinear Dynamics (La Jolla Institute)	78-57870	0-88318-145-2
No. 47	Clustering Aspects of Nuclear Structure and Nuclear Reactions (Winnepeg, 1978)	78-64942	0-88318-146-0
No. 48	Current Trends in the Theory of Fields (Tallahassee, 1978)	78-72948	0-88318-147-9
No. 49	Cosmic Rays and Particle Physics – 1978 (Bartol Conference)	79-50489	0-88318-148-7
No. 50	Laser-Solid Interactions and Laser Processing – 1978 (Boston)	79-51564	0-88318-149-5
No. 51	High Energy Physics with Polarized Beams and Polarized Targets (Argonne, 1978)	79-64565	0-88318-150-9
No. 52	Long-Distance Neutrino Detection – 1978 (C.L. Cowan Memorial Symposium)	79-52078	0-88318-151-7
No. 53	Modulated Structures – 1979 (Kailua Kona, Hawaii)	79-53846	0-88318-152-5
No. 54	Meson-Nuclear Physics – 1979 (Houston)	79-53978	0-88318-153-3
No. 55	Quantum Chromodynamics (La Jolla, 1978)	79-54969	0-88318-154-1
No. 56	Particle Acceleration Mechanisms in Astrophysics (La Jolla, 1979)	79-55844	0-88318-155-X
No. 57	Nonlinear Dynamics and the Beam-Beam Interaction (Brookhaven, 1979)	79-57341	0-88318-156-8
No. 58	Inhomogeneous Superconductors – 1979 (Berkeley Springs, W.V.)	79-57620	0-88318-157-6
No. 59	Particles and Fields – 1979 (APS/DPF Montreal)	80-66631	0-88318-158-4
No. 60	History of the ZGS (Argonne, 1979)	80-67694	0-88318-159-2
No. 61	Aspects of the Kinetics and Dynamics of Surface Reactions (La Jolla Institute, 1979)	80-68004	0-88318-160-6
No. 62	High Energy e^+e^- Interactions (Vanderbilt, 1980)	80-53377	0-88318-161-4
No. 63	Supernovae Spectra (La Jolla, 1980)	80-70019	0-88318-162-2
No. 64	Laboratory EXAFS Facilities – 1980 (Univ. of Washington)	80-70579	0-88318-163-0
No. 65	Optics in Four Dimensions – 1980 (ICO, Ensenada)	80-70771	0-88318-164-9
No. 66	Physics in the Automotive Industry – 1980 (APS/AAPT Topical Conference)	80-70987	0-88318-165-7
No. 67	Experimental Meson Spectroscopy – 1980 (Sixth International Conference, Brookhaven)	80-71123	0-88318-166-5
No. 68	High Energy Physics – 1980 (XX International Conference, Madison)	81-65032	0-88318-167-3
No. 69	Polarization Phenomena in Nuclear Physics – 1980 (Fifth International Symposium, Santa Fe)	81-65107	0-88318-168-1

No. 70	Chemistry and Physics of Coal Utilization – 1980 (APS, Morgantown)	81-65106	0-88318-169-X
No. 71	Group Theory and its Applications in Physics – 1980 (Latin American School of Physics, Mexico City)	81-66132	0-88318-170-3
No. 72	Weak Interactions as a Probe of Unification (Virginia Polytechnic Institute – 1980)	81-67184	0-88318-171-1
No. 73	Tetrahedrally Bonded Amorphous Semiconductors (Carefree, Arizona, 1981)	81-67419	0-88318-172-X
No. 74	Perturbative Quantum Chromodynamics (Tallahassee, 1981)	81-70372	0-88318-173-8
No. 75	Low Energy X-Ray Diagnostics – 1981 (Monterey)	81-69841	0-88318-174-6
No. 76	Nonlinear Properties of Internal Waves (La Jolla Institute, 1981)	81-71062	0-88318-175-4
No. 77	Gamma Ray Transients and Related Astrophysical Phenomena (La Jolla Institute, 1981)	81-71543	0-88318-176-2
No. 78	Shock Waves in Condensed Mater – 1981 (Menlo Park)	82-70014	0-88318-177-0
No. 79	Pion Production and Absorption in Nuclei – 1981 (Indiana University Cyclotron Facility)	82-70678	0-88318-178-9
No. 80	Polarized Proton Ion Sources (Ann Arbor, 1981)	82-71025	0-88318-179-7
No. 81	Particles and Fields –1981: Testing the Standard Model (APS/DPF, Santa Cruz)	82-71156	0-88318-180-0
No. 82	Interpretation of Climate and Photochemical Models, Ozone and Temperature Measurements (La Jolla Institute, 1981)	82-71345	0-88318-181-9
No. 83	The Galactic Center (Cal. Inst. of Tech., 1982)	82-71635	0-88318-182-7
No. 84	Physics in the Steel Industry (APS/AISI, Lehigh University, 1981)	82-72033	0-88318-183-5
No. 85	Proton-Antiproton Collider Physics –1981 (Madison, Wisconsin)	82-72141	0-88318-184-3
No. 86	Momentum Wave Functions – 1982 (Adelaide, Australia)	82-72375	0-88318-185-1
No. 87	Physics of High Energy Particle Accelerators (Fermilab Summer School, 1981)	82-72421	0-88318-186-X
No. 88	Mathematical Methods in Hydrodynamics and Integrability in Dynamical Systems (La Jolla Institute, 1981)	82-72462	0-88318-187-8
No. 89	Neutron Scattering – 1981 (Argonne National Laboratory)	82-73094	0-88318-188-6
No. 90	Laser Techniques for Extreme Ultraviolt Spectroscopy (Boulder, 1982)	82-73205	0-88318-189-4

No. 91	Laser Acceleration of Particles (Los Alamos, 1982)	82-73361	0-88318-190-8
No. 92	The State of Particle Accelerators and High Energy Physics (Fermilab, 1981)	82-73861	0-88318-191-6
No. 93	Novel Results in Particle Physics (Vanderbilt, 1982)	82-73954	0-88318-192-4
No. 94	X-Ray and Atomic Inner-Shell Physics – 1982 (International Conference, U. of Oregon)	82-74075	0-88318-193-2
No. 95	High Energy Spin Physics – 1982 (Brookhaven National Laboratory)	83-70154	0-88318-194-0
No. 96	Science Underground (Los Alamos, 1982)	83-70377	0-88318-195-9
No. 97	The Interaction Between Medium Energy Nucleons in Nuclei – 1982 (Indiana University)	83-70649	0-88318-196-7
No. 98	Particles and Fields – 1982 (APS/DPF University of Maryland)	83-70807	0-88318-197-5
No. 99	Neutrino Mass and Gauge Structure of Weak Interactions (Telemark, 1982)	83-71072	0-88318-198-3
No. 100	Excimer Lasers – 1983 (OSA, Lake Tahoe, Nevada)	83-71437	0-88318-199-1
No. 101	Positron-Electron Pairs in Astrophysics (Goddard Space Flight Center, 1983)	83-71926	0-88318-200-9
No. 102	Intense Medium Energy Sources of Strangeness (UC-Sant Cruz, 1983)	83-72261	0-88318-201-7
No. 103	Quantum Fluids and Solids – 1983 (Sanibel Island, Florida)	83-72440	0-88318-202-5
No. 104	Physics, Technology and the Nuclear Arms Race (APS Baltimore –1983)	83-72533	0-88318-203-3
No. 105	Physics of High Energy Particle Accelerators (SLAC Summer School, 1982)	83-72986	0-88318-304-8
No. 106	Predictability of Fluid Motions (La Jolla Institute, 1983)	83-73641	0-88318-305-6
No. 107	Physics and Chemistry of Porous Media (Schlumberger-Doll Research, 1983)	83-73640	0-88318-306-4
No. 108	The Time Projection Chamber (TRIUMF, Vancouver, 1983)	83-83445	0-88318-307-2
No. 109	Random Walks and Their Applications in the Physical and Biological Sciences (NBS/La Jolla Institute, 1982)	84-70208	0-88318-308-0
No. 110	Hadron Substructure in Nuclear Physics (Indiana University, 1983)	84-70165	0-88318-309-9
No. 111	Production and Neutralization of Negative Ions and Beams (3rd Int'l Symposium, Brookhaven, 1983)	84-70379	0-88318-310-2

No. 112	Particles and Fields – 1983 (APS/DPF, Blacksburg, VA)	84-70378	0-88318-311-0
No. 113	Experimental Meson Spectroscopy – 1983 (Seventh International Conference, Brookhaven)	84-70910	0-88318-312-9
No. 114	Low Energy Tests of Conservation Laws in Particle Physics (Blacksburg, VA, 1983)	84-71157	0-88318-313-7
No. 115	High Energy Transients in Astrophysics (Santa Cruz, CA, 1983)	84-71205	0-88318-314-5
No. 116	Problems in Unification and Supergravity (La Jolla Institute, 1983)	84-71246	0-88318-315-3
No. 117	Polarized Proton Ion Sources (TRIUMF, Vancouver, 1983)	84-71235	0-88318-316-1
No. 118	Free Electron Generation of Extreme Ultraviolet Coherent Radiation (Brookhaven/OSA, 1983)	84-71539	0-88318-317-X
No. 119	Laser Techniques in the Extreme Ultraviolet (OSA, Boulder, Colorado, 1984)	84-72128	0-88318-318-8
No. 120	Optical Effects in Amorphous Semiconductors (Snowbird, Utah, 1984)	84-72419	0-88318-319-6
No. 121	High Energy e^+e^- Interactions (Vanderbilt, 1984)	84-72632	0-88318-320-X
No. 122	The Physics of VLSI (Xerox, Palo Alto, 1984)	84-72729	0-88318-321-8
No. 123	Intersections Between Particle and Nuclear Physics (Steamboat Springs, 1984)	84-72790	0-88318-322-6
No. 124	Neutron-Nucleus Collisions – A Probe of Nuclear Structure (Burr Oak State Park - 1984)	84-73216	0-88318-323-4
No. 125	Capture Gamma-Ray Spectroscopy and Related Topics – 1984 (Internat. Symposium, Knoxville)	84-73303	0-88318-324-2
No. 126	Solar Neutrinos and Neutrino Astronomy (Homestake, 1984)	84-63143	0-88318-325-0
No. 127	Physics of High Energy Particle Accelerators (BNL/SUNY Summer School, 1983)	85-70057	0-88318-326-9
No. 128	Nuclear Physics with Stored, Cooled Beams (McCormick's Creek State Park, Indiana, 1984)	85-71167	0-88318-327-7
No. 129	Radiofrequency Plasma Heating (Sixth Topical Conference, Callaway Gardens, GA, 1985)	85-48027	0-88318-328-5
No. 130	Laser Acceleration of Particles (Malibu, California, 1985)	85-48028	0-88318-329-3
No. 131	Workshop on Polarized ^{3}He Beams and Targets (Princeton, New Jersey, 1984)	85-48026	0-88318-330-7
No. 132	Hadron Spectroscopy–1985 (International Conference, Univ. of Maryland)	85-72537	0-88318-331-5

No. 133	Hadronic Probes and Nuclear Interactions (Arizona State University, 1985)	85-72638	0-88318-332-3
No. 134	The State of High Energy Physics (BNL/SUNY Summer School, 1983)	85-73170	0-88318-333-1
No. 135	Energy Sources: Conservation and Renewables (APS, Washington, DC, 1985)	85-73019	0-88318-334-X
No. 136	Atomic Theory Workshop on Relativistic and QED Effects in Heavy Atoms	85-73790	0-88318-335-8
No. 137	Polymer-Flow Interaction (La Jolla Institute, 1985)	85-73915	0-88318-336-6
No. 138	Frontiers in Electronic Materials and Processing (Houston, TX, 1985)	86-70108	0-88318-337-4